Comparative Pathobiology

Volume 5

STRUCTURE OF MEMBRANES AND RECEPTORS

Comparative Pathobiology

Volume 1: Biology of the Microsporidia
Volume 2: Systematics of the Microsporidia
Volume 3: Invertebrate Immune Responses
Volume 4: Invertebrate Models for Biomedical Research
Volume 5: Structure of Membranes and Receptors

Comparative Pathobiology

Volume 5

STRUCTURE OF MEMBRANES AND RECEPTORS

Edited by Thomas C. Cheng
Medical University of South Carolina
Charleston, South Carolina

Plenum Press • New York and London

Library of Congress Cataloging in Publication Data

Main entry under title:

Structure of membranes and receptors.

(Comparative pathobiology; v. 5)
"Proceedings of a special symposium on structure of membranes and receptors, held February 19–20, 1981 at the Medical University of South Carolina, Charleston, South Carolina."
Includes bibliographical references and index.
1. Cell membranes—Congresses. 2. Cell receptors—Congresses. 3. Immunogenetics—Congresses. I. Cheng, Thomas Clement. II. Series.
QH601.S778 1984 574.87′5 83-23780
ISBN 0-306-41503-8

Proceedings of a special symposium on Structure of Membranes and Receptors, held February 19–20, 1981, at the Medical University of South Carolina, Charleston, South Carolina

A Division of Plenum Publishing Corporation
233 Spring Street, New York, N.Y. 10013

Printed in the United States of America

PREFACE

On February 19 and 20, 1981, a special symposium was held at the Medical University of South Carolina in Charleston. The common theme of this series of presentations was "Structure of Membranes and Receptors." Although most of the papers were by members of the faculty of this institution, through a grant from the office of Dr. John W. Zemp, Dean of the College of Graduate Studies, we were fortunate to have had with us Dr. Lee Hood of the California Institute of Technology, Dr. Michael J. Crumpton of the Imperial Cancer Research Fund, United Kingdom, and Dr. David D. Sabatini of the New York University Medical Center.

This symposium, for which the following pages represent the proceedings, was organized by Dr. John J. Marchalonis of the Department of Biochemistry and Dr. Edward L. Hogan of the Department of Neurology, both of the Medical University of South Carolina.

The common theme is a timely one since, as any reader will recognize, the structure of membranes and receptors, whether pertaining to cellular immunology, developmental biology, microbiology, cellular endocrinology, or experimental neurology, represents a major cutting edge of modern biology today. The organization and participants of the symposium enthusiastically endorsed the idea that what was presented in Charleston, South Carolina, should be shared with colleagues around the world. Consequently, plans were made to publish these proceedings as Volume 5 of _Comparative Pathobiology_.

The editor wishes to take this opportunity to announce that Dr. Lee A. Bulla, Jr. the former co-editor of _Comparative Pathobiology_, has elected to retire from this responsibility, leaving me, at least for the time being, as the sole editor. The original idea that stimulated the launching of this series by Plenum Press, New York and London, however, remains unaltered.

Comparative Pathobiology is intended for the publication of the proceedings of special symposia dealing with all aspects of what is commonly designated as pathobiology that the editor deems worthwhile sharing with the scientific community. Those interested in supporting this objective should contact the editor.

Thomas C. Cheng
Charleston, South Carolina

CONTENTS

Section I - Structure of Membranes and Receptors

Section II - Structure and Organization of Genes

IMMUNOGLOBULIN-RELATED ANTIGEN-RECEPTORS OF THYMUS-DERIVED LYMPHOCYTES[1]

John J. Marchalonis

Department of Biochemistry
Medical University of North Carolina
171 Ashley Avenue
Charleston, South Carolina 29425

I. INTRODUCTION

Lymphocytes are small round cells of rather unprepossessing appearance, yet they possess a potential for recognition of foreignness and for differentiation into effector cells that makes them a prominent bulwark of the body's defenses. Two broad functional classes of lymphocytes have been described and studied with great vigor during the past 15 years. These are the B cells, or bone-marrow-derived lymphocytes, which are the precursors of antibody forming cells, and the so-called T cells for thymus-derived lymphocytes. T cells play the major role in cell mediated immunity, e.g., the rejection of tumors and allografts, and are required for collaboration with B cells in the elaboration of immune responses

1. Supported in part by grant RD-101 from the American Cancer Society and 1R01/AI17493-01 from the National Institutes of Health.

to many antigens (Marchalonis, 1977a). The mechanisms by which T cells interact with one another, with nonlymphoid accessory cells and with B cells are extremely complicated and various models have been proposed for the regulation of an immunological network. I will not consider the mechanisms here but will focus upon the molecular properties of the primary receptor for antigen as it occurs on the surface of antigen-specific T cells. It is clear at this time that in both man and mouse the antigen receptor on B cells resembles serum immunoglobulin and bears not only immunoglobulin combining site regions but class specific determinants of the IgM and IgD isotypes (Vitetta and Uhr, 1975). The problem of identifying and characterizing the antigen-specific receptor on T cells has been a controversial and challenging issue in cellular and molecular immunology since the days when it was first recognized that T cells could interact specifically with antigen. The minimal hypothesis regarding this receptor is that the T cell surface should possess combining site similar or identical to those immunoglobulin variable (V) regions but its constant regions are not necessarily identical to those of circulating antibodies. This hypothesis is illustrated in Fig. 1 which compares a B cell IgM receptor with circulating IgG antibody to the same antigen. In order to test this hypothesis, it was necessary to derive a battery of antisera directed against combining site region determinants of serum antibodies and to develop means for detecting whether such antibodies reacted with T cells and with T cell products. In addition, it was necessary to develop a sensitive radiochemical method for labeling external surface proteins of plasma membranes and subsequently using the reactive antibodies as probes for the isolation of the surface molecules (Marchalonis et al., 1971). I will present data supporting the conclusion that certain, antigen-specific T cells express and produce molecules related, but not identical, to serum immunoglobulins and will describe some biochemical characterization data comparing the T cell immunoglobulin-like receptor with classical serum antibodies.

II. EXPERIMENTAL EVIDENCE

We have used two approaches to the problem of plasma membrane receptors of T and B cells. One approach was to use the enzyme lactoperoxidase as a catalyst to incorporate carrier-free ^{125}I-iodide into tyrosyl residues of exposed protein on the external face of the plasma membrane (Marchalonis et al., 1971). This approach provided sufficient sensitivity to enable us to carry out microanalytical studies on the serological and biochemical properties of receptors isolated from a few as 2×10^6 T cells. The other approach which we used was based upon the fractionation procedures for membranes devised by Crumpton and Snary (1974) which entailed disruption of the cells, starting with approximately 10^{10} or more lymphocytes, and isolation of the plasma membrane by differential

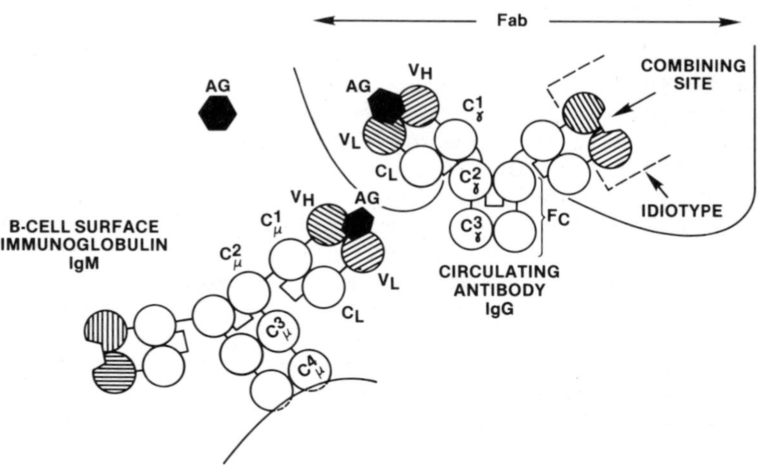

Fig. 1. Diagram illustrating the similarities and differences between B cell surface IgM receptors for a given antigen and the circulating IgG antibody to that antigen. Ag antigen; Fab, antigen binding fragment; Fc, Fc portion of IgG immunoglobulin comprised of the second and third constant region domains. The Fc fragment of the surface μ chain would be comprised of constant region domains of the μ chain. Although the circulating antibody and the surface IgM have the same combining site for antigen they represent different classes as defined by their heavy chains. Furthermore, surface μ chain has a hydrophobic tail which allows it to intercalate with the plasma membrane. It differs in this structure from the μ chain of circulating IgM. The IgM plasma membrane molecule is also a 7S structure composed of two Lμ pairs, rather than the cyclic pentamer characteristic of secreted antibody. Idiotypic determinants are combining site region-related determinants and are usually formed by interaction between the V_H and V_L portions of the molecules, although some idiotypic determinants are specified predominantly by the V_H structure.

centrifugation. As can be seen in Fig. 2 which presents a stained pattern of the plasma membrane proteins of the T cell lymphoma WEHI 22 as resolved on polyacrylamide gel electrophoresis in sodium dodecyl sulfate, more than 50 bands can be detected. Because of the complexity of the patterns, it was necessary to use affinity techniques to isolate particular components. We have used three types of approaches: In the first approach we used affinity chromatography on insolubilized lectins in order to isolate glycosylated membrane proteins bearing terminal sugars of particular sequences (Hunt and Marchalonis, 1974); the second approach was to use insolubilized antibodies as our immune affinity reagent (Hunt and Marchalonis, 1974); and in the third approach, we isolated ligand-binding surface proteins using insolubilized ligand such as the hapten dinitrophenyl coupled to Sepharose (Marchalonis, 1976).

The great selective power of immune affinity chromatography is illustrated in Fig. 3 which presents an SDS-polyacrylamide gel of radioiodinated surface immunoglobulins of murine B cells which were isolated in a single step by binding to rabbit antibody to mouse κ light chain coupled to Sepharose 4B. Only three components can be detected in this autoradiograph. One is the μ chain of the cell surface 7S IgM; another is the δ chain of cell surface IgD which can appear on the same cell; and the third component is the light chain (κ), which can occur in covalent association with either the μ or δ chain. The δ chain envelope is broad relative to that of μ chain, and this is because the δ chain from the cell surface shows considerable heterogeneity, possibly due either to differing degrees of glycosylation or to proteolysis.

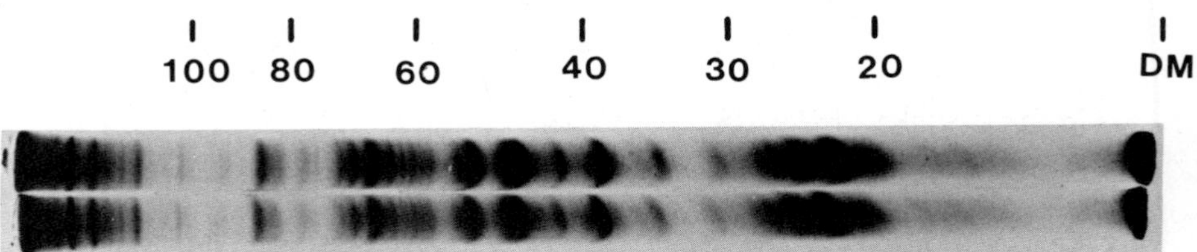

Fig. 2. Resolution by polyacrylamide gel electrophoresis in sodium dodecyl-containing buffers in the presence of mercaptoethanol of isolated plasma membranes of the murine T cell lymphoma WEHI 22. DM, dyemaker. The numbers indicate the position at which proteins of known molecular weight migrate under these gel conditions (MW x 1000).

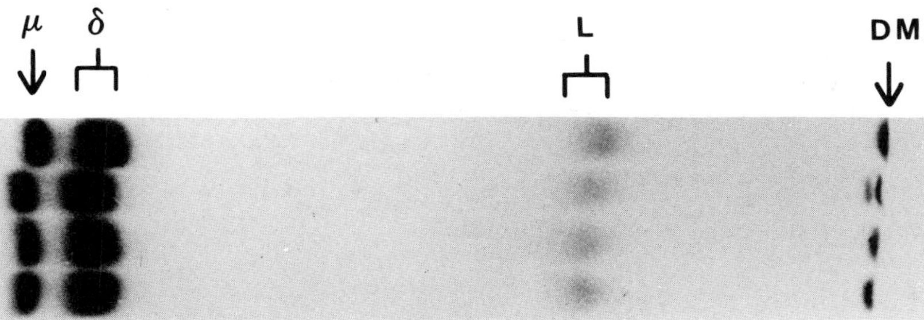

Fig. 3. Radioautograph of ^{125}I-labeled surface immunoglobulins of murine B cells which were surface-radioiodinated using the lactoperoxidase catalyzed technique (Marchalonis et al., 1971) and isolated as a single step by immune affinity chromatography using insolubilized antibody to mouse κ chain. Four replicate samples were resolved by polyacrylamide gel electrophoresis in sodium dodecylsulfate-containing buffers in the presence of mercaptoethanol. DM, dyemaker. L indicates the position at which light chains migrate; μ and δ refer to the B cell surface μ and δ chains. B cell 7S IgM exists in the form $(\kappa\mu)_2$; cell surface IgD exists in the form $(\kappa\delta)_2$.

In studies of the T cell receptor, we and others found that antisera which were specific for the constant regions of the heavy chains or light chains, i.e., the antibodies which reacted with the dominant isotype-specific determinants, were not reactive with the T cell product. However, it was found that certain antiserum made against normal, polyclonal serum immunoglobulins did react with T cell products (Marchalonis et al., 1972; Burckhardt et al., 1974), and, moreover, antibodies made against individually specific determinants (idiotypes) of antibodies or homogenous myeloma proteins did react with T cell products (Binz and Wigzell, 1977; Rajewsky and Eichmann, 1977).

Table 1 lists the type of antibodies which in our hands were found to react with T cell receptors for antigen. In essence, the first four types of antibodies are ones which have been raised against the combining site region either of polyclonal, normal immunoglobulins in the case of the chicken antibodies against Fab interaction determinant or reactive with combining site determinants of particular antibodies. See Fig. 1 above for graphic representation of the location of these fragments. In our case, we raised rabbit antibodies against the combining site of murine antibodies

Table 1. Antibodies which react with T cell receptors for antigen.

(1) Anti-fab interaction determinants (raised in chickens)

(2) Anti-V_H fragment of μ chain (raised in chickens)

(3) Anti-Fdμ fragment

(4) Anti-idiotype

(5) Putative Anti-TgT

from strain A/J mice which reacted with the arsonate hapten (Tung and Nisonoff, 1975; Marchalonis et al., 1979). Antibodies of type 1 which are directed against a polyclonal pool of Fab interaction determinants can be termed "shotguns" because they react with the vast majority of the pool of Fab interaction determinants as it occurs in serum and on cells. By contrast, the so-called anti-idiotypes are very specific and resemble antigen in their selectivity in that they only react with a particular subset of antibody combining sites which has light and heavy chains of the proper sequence. The Fab fragment is formed of light chain plus the so-called Fd piece of the heavy chain. This fragment contains the combining site for antigen which is formed by interaction between the variable region of the light chain and the variable region of the heavy chain. We have produced these types of antisera against immunoglobulins and isolated T cell products of murine and human, as well as other primates.

It is possible to use these anti-combining site reagents to detect immunoglobulin related components on the surface of certain T cells of normal individuals and on certain *in vitro* grown T cell leukemia or lymphoma lines. Fig. 4 illustrates immunocytofluorescence data comparing the binding of chicken anti-mouse Fab (shotgun) developed with fluorescein-labeled rabbit anti-chicken immunoglobulin (B,D) and the binding by one of the cells of rodamine labeled myoglobin (A) or horse spleen ferritin (C). The usual frequency for binding of protein antigens by T cells of unimmunized animals is about one per thousand T cells (DeLuca et al., 1979). It is an important observation here that the distribution of antigen on the cell surface in A and B and C and D is the same as the distribution of the immunoglobulin. This result is expected if the immunoglobulin serves as the primary binding receptor for the antigen. We have carried out a series of studies in which there was an absolute correspondence between antigen Fab-related determinant using a variety of antigens and counting close to fifty individual cells (DeLuca et al., 1979). This sort of observation provides strong presumptive evidence that the immunoglobulin-like molecule

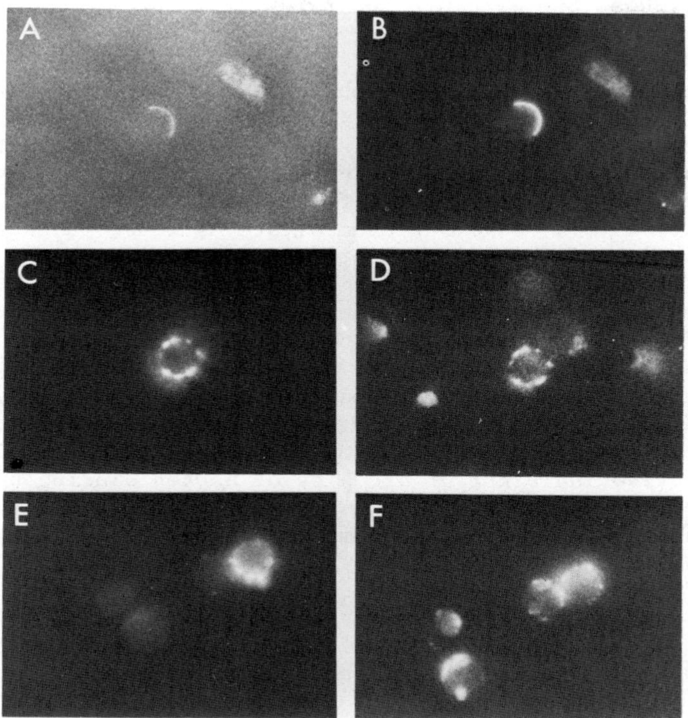

Fig. 4. Codistribution of rodamine-labeled horse spleen ferritin (A) and chicken anti-mouse Fab plus fluorescein-labeled rabbit anti-chicken immunoglobulin (B) on a thymic antigen binding cell. Codistribution of rodamine-labeled keyhole limpit hemocyanin (C) and chicken anti-mouse Fab plus fluorescein-labeled rabbit antibodies to chicken IgG (D) on a second thymic antigen binding T cell. Photographs E and F show the lack of codistribution of spots of rodamine-labeled keyhole limpit hemocyanin and antibody to histocompatibility antigen ($H\text{-}2^d$) plus fluorescein-labeled rabbit anti-mouse immunoglobulin respectively. Since the anti-$H\text{-}2^d$ is detected by fluorescein-labeled rabbit anti-mouse Ig, the lack of codistribution shown in E and F also indicates that the thymic antigen binding cells are T cells rather than B cells (x1500). (Data of DeLuca et al., 1979).

is the primary receptor for antigen on T cells. By contrast in control experiments, neither the Thy-1 alloantigen nor the H-2 histocompatibility antigens codistributed with foreign antigens (E and F). Thus, the evidence indicates that these molecules are not primary receptors for antigen.

We further approached the problem of specificity by comparing our rabbit anti-idiotype with our shotgun chicken anti-Fab for their capacities to inhibit the binding of serum antibody to the arsonate hapten and also the binding of the arsonate hapten by antigen specific T cells isolated from immunized strain A/J mice (Warr et al., 1979). Fig. 5 presents data obtained in experiments blocking binding of the idiotype bearing antibodies to the hapten in solution. It can be seen that the binding of hapten by the combining site of the idiotype-bearing molecule is blocked by anti-idiotype but not by normal rabbit serum or by rabbit antiserum directed against the constant region of the heavy chain (anti-IgG 2). This is the expected result and confirms that our anti-idiotype, which is directed against the V_H/V_L interaction determinant, reacts with the combining site. In addition, our "shotgun" anti-Fab also is a very effective inhibitor of blocking of hapten binding by the idiotype-bearing molecule. These data do not indicate that the "shotgun" and the anti-idiotype compete for precisely the same determinants on the antibody. They do, however, establish that both compete for the same general region of the molecule. The parallel experiment was performed using antigen-binding T cells in which the arsonate hapten was coupled to fluorescein-labeled bovine serum albumin and the assay was performed by cytofluorescence. Approximately 2-4% of peripheral T lymphocytes of the immunized A/J mice bound the antigen specifically as shown in Fig. 6. The specificity of the binding for the arsonate hapten is shown by the virtually complete inhibition by arsonate-derivatized bovine serum albumin coupled with the fact that bovine serum albumin alone does not inhibit binding at all. The rabbit anti-idiotype showed substantial blocking up to approximately 80%, a value which is expected because approximately 70-80% of antibodies to the arsonate hapten bear the crossreactive idiotype (Ju et al., 1977). The chicken anti-Fab gave total inhibition of the reaction, and the specificity of this is shown because it was possible to inhibit the blocking by titrating in normal polyclonal mouse immunoglobulin which bore the collection of Fab region determinants. This evidence, thus, indicates that antigen specific T cells express an Fab-related surface molecule which bears immunoglobulin idiotype and is also detectable using chicken antibodies directed against combining site determinants found on pooled murine immunoglobulin.

Is the Ig-related T cell product identical to serum antibodies or to B cell surface immunoglobulins? This has been an extremely contentious issue and many papers have been written disputing this

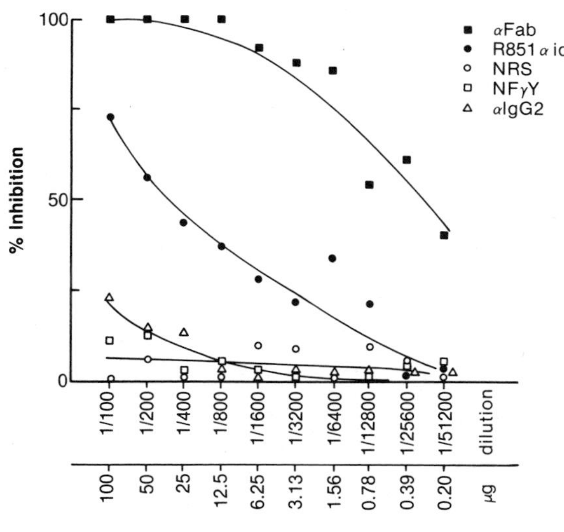

Fig. 5. Inhibition of the binding of idiotype-bearing anti-ARS antibody to the ARS hapten. Anti- (Fab), chicken antibody to the $(Fab')_2$ fragment of mouse IgG; R815, anti-idiotypic antiserum to the idiotypic determinants of the murine anti-arsonate (ARS) antibody; NRS, normal rabbit serum; anti-IgG_2, rabbit anti-serum to mouse IgG_{2a} and IgG_{2b} (these are the class specific constant region determinants of the idiotype bearing IgG antibodies studied here); NFγY, normal chicken IgY. The dilution scale refers to the intact sera, the microgram scale refers to the anti-$(Fab')_2$ and the normal chicken IgY, both of which are purified immunoglobulins. (Data of Warr et al., 1979).

point. The most direct way to approach the problem is to isolate the T cell product and to compare them directly with the B cell immunoglobulins by serological and biochemical techniques. When this is done, it is clear that the answer to this question is "no". Fig. 7 illustrates the idiotype-positive molecule synthesized by peripheral T cells of A/J mice immunized to the arsonate hapten. The only significant component detectable has an approximate mass of 67,000 daltons. The peripheral T cells were stimulated by the T cell mitogen Concanavalin A prior to pulse labeling with ^{75}Se-methionine. A similar preparation utilizing purified B cells from the same animal was negative for the presence of componants in the 67,000 dalton range. The existence of a T cell derived molecule which bears antigenic determinants related to those of

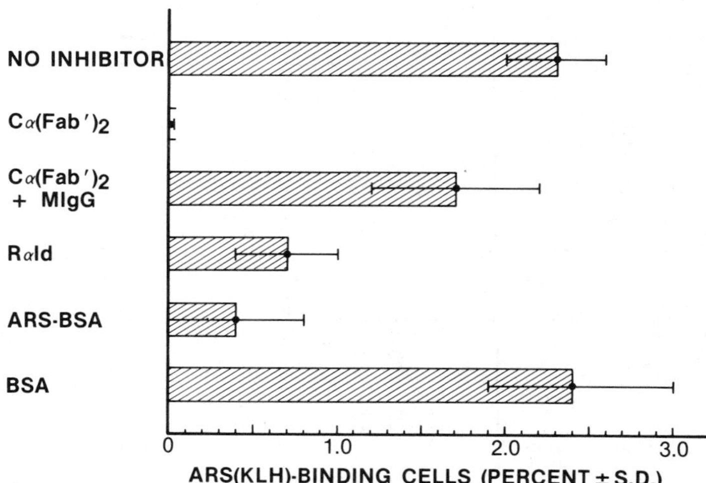

Fig. 6. Inhibition of binding of the arsonate (ARS) hapten by peripheral T cells of primed A/J mice by anti-idiotype and by chicken anti-Fab reagents. The arsonate hapten was coupled to rodamine-labeled keyhole limpit hemocyanin for the use in these binding studies. The specificity of the reaction for the arsonate hapten is illustrated by the fact that ARS coupled to bovine serum albumin (ARS-BSA) inhibits binding whereas the BSA carrier alone does not. The reaction is blocked to about 80% by rabbit anti-idiotye (RαID) and is completely inhibited with the chicken anti-mouse Fab fragment. Addition of purified murine IgG immunoglobulin (MIgG) reverses the inhibition by the chicken anti-Fab, thereby indicating the specificity of the inhibition reaction. (Based upon data of Warr et al., 1979).

immunoglobulin combining sites is further illustrated in Fig. 8 which presents a fluorograph obtained for biosynthesis of a V_H-bearing molecule by a long-term marmoset amplifier T cell line 70-N2. Although some material is present in the control, and most probably represents actin or proteins which bind to the Fc fragment of immunoglobulin. It is noteworthy that appreciable amounts of material resembling light polypeptide chains in SDS gel mobility are absent in both the mouse and the primate T cell biosynthetically labeled product (Marchalonis et al., 1980b). In some cases, we and others have observed polypeptide chains resembling light chains to

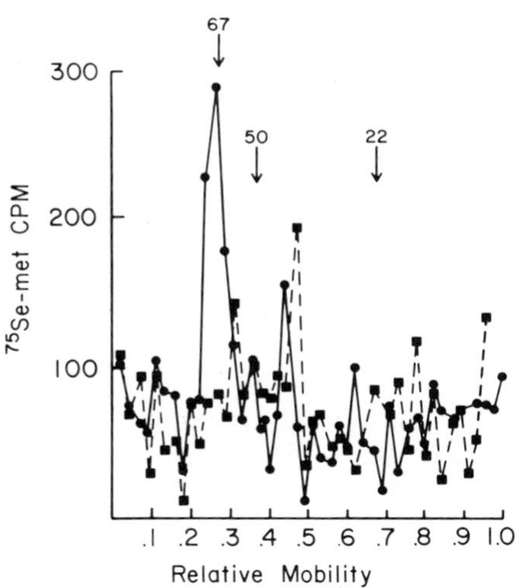

Fig. 7. Resolution by polyacrylamide gel electrophoresis in sodium dodecylsulfate-containing buffers of biosynthetically labeled (^{75}Se-methionine) molecule which bears the anti-ARS idiotype produced by A/J T cells from animals primed with the arsonate hapten. ●——●, idiotype-bearing molecule; ■----■ , control in which the same anti-idiotype sera were used to investigate B cell isolated from the same animal. Both purified T and B cell preparations were stimulated with the mitogen Concanavalin A. Figure courtesy of Drs. D. DeLuca and J. M. Decker.

be associated with the 67,000 dalton heavy chains. The appearance of the light chain-like components is not as reliable, however, as is the presence of the heavy chain-like material. The heavy chain-like material can associate into disulfide bonded dimers of approximately 140,000 molecular weight (Warr et al., 1978; Marchalonis and Wang, 1981).

Functional studies indicated that the immunoglobulin-related product of T cells bound cytophilically to macrophages, whereas surface IgM and IgD of B cells did not (Cone et al., 1974). This functional difference suggested that the T and B cell products must be distinct, and this difference was supported by more recent serological and biochemical studies (Cone, 1981; Binz and Wigzell, 1977; Rajewsky and Eichmann, 1977; Marchalonis, 1980c; Moseley et al., 1979). The T cell product was found to express an antigenic determinant which crossreacts with, but is not identical to, a determinant which occurs in the Fd fragment of μ chain (Marchalonis

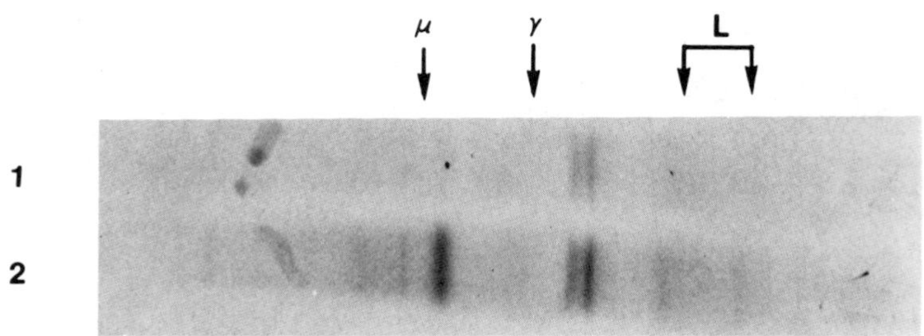

Fig. 8. Autoradiography of biosynthetically-labeled V_H-bearing molecule synthesized by a primate amplifier T cell line. The cell line used with 70-N2 which was derived from the cotton-topped marmoset. This cell line has the surface markers and functional activities of a human amplifier T cell. 1, Material precipitated using normal chicken serum and rabbit antibodies directed against normal chicken globulin. Lane 2, material precipitated using chicken antibodies to the human V_H fragment followed by rabbit antibodies to chicken immunoglobulin. The cells were biosynthetically-labeled using tritiated leucine. Counts were precipitated from the culture fluid following 6 hours of incubation. μ, γ, and L represent the positions at which immunoglobulin μ chain, γ chain and light chains migrate, respectively. (From Marchalonis et al., 1980b).

et al., 1980b). Antisera against major isotype-specific determinants of serum antibodies do not react with the T cell product (Marchalonis, 1975; Cone, 1981). The general similarity, but lack of identity, between the T cell heavy chain and serum μ is indicated in Fig. 9, which is an autoradiograph of iodine-labeled, tyrosine containing tryptic peptides of T cell heavy chain (A), μ chain of the myeloma protein MOPC 104E (B), and the viral glycoprotein in gp 71 (C). The μ chain and the τ chain share some peptide clusters, but are distinct in others, whereas the gp 71 profile is totally distinct from that of either the μ chain or the τ chain (Moseley et al., 1979). It has proven possible to isolate CNBr-produced peptides from the Fd fragment of the primate τ chain, and sequencing studies are in progress to determine the exact relationship between this molecule and the Fd fragments of serum immunoglobulins.

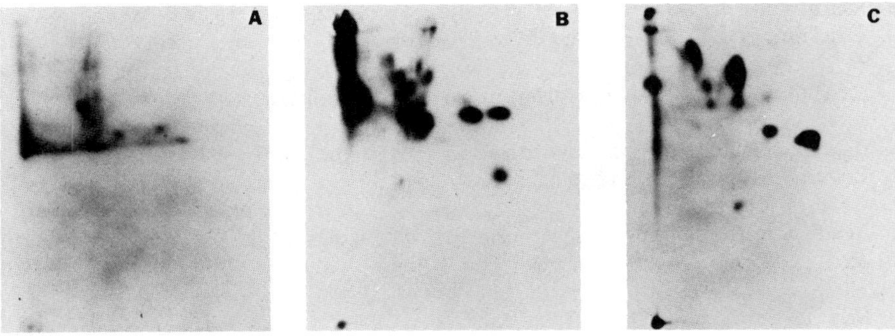

Fig. 9. Resolution by two dimensional peptide mapping of tyrosine-containing tryptic peptides of A, heavy chain of the immunoglobulin like molecule produced by the monoclonal murine T cell line WEHI 7; B, the μ chain of the murine myeloma protein MOPC 104E; C, the viral glycoprotein Gp 71. Tyrosines were labeled with ^{125}I and the pattern were detected by autoradiography. (Data of Moseley et al., 1979).

III. CONCLUSIONS

Our data, as well as those from other laboratories (Binz and Wigzell, 1977; Rajewsky and Eichmann, 1977; Pacifico and Capra, 1980), now strongly support the conclusion that some T cells express and synthesize a surface component which bears antigenic determinants related to those found in the combining site region of serum antibodies, but that the T cell molecules express constant regions which are distinct from any of the major serum isotypes. The properties of the T cell molecule are outlined in Table 2. Although various workers have now observed structural and serological similarities between the τ chain and the μ chain (Burckhardt et al., 1974; Hammerling et al., 1976; Baird, 1979; Marchalonis et al., 1980c: Moroz and Hahn, 1973), we do not believe that the τ chain is a particular subclass of μ chain. Rather, we belive that the T cell receptor for heavy chain derives directly from the ancestral immunoglobulin heavy chain in evolution which would resemble a μ chain (Marchalonis, 1977b). Present evidence indicates that the μ chain has been conserved throughout vertebrate evolution, and it is reasonable to expect that the T cell heavy chain would

Table 2. Properties of Ig-Related T cell Receptor

(1) Expresses variable regions antigenically related to Ig variable regions (V_H definite; V_L questionable).

(2) Codistributes with antigen on antigen-binding T cells.

(3) Contains heavy chains apparent M.W. 65,000 - 70,000, which can form S-S bonded dimers.

(4) Sometimes contain light chains of approx. M.W. 23,000 which are not covalently linked to heavy chains.

(5) Not a serum or B cell-associated isotype.

(6) Heavy chains share peptides with μ and α chains.

also be conserved because the molecule as it exists in mammals serves essentially the same functions as does the molecule in primitive vertebrates. For this reason, we would expect that both the μ chain and the τ chain would be related to the ancestral immunoglobulin heavy chain, and, therefore, to show a relationship to one another. It is likely that a family of T cell isotypes exists; for example, possibly suppressor T cells would express one constant region whereas helper T cells would express another, although both would use variable regions obtained from the pool of immunoglobulin variable region.

The question remains to be determined whether the variable regions of T cell receptors are representative of those expressed on serum or B cell immunoglobulins, and this question must be answered by amino acid sequence analysis. In addition, final appraisal of the degree of similarity or difference between the constant regions of T cell receptors and those of their serum or B cell homologues likewise requires the attainment of detailed amino acid sequence data.

ACKNOWLEDGEMENT

I thank Ms. Mary A. Jackson for typing the manuscript.

IV. REFERENCES

Baird, S. (1980). The lymphoid tissue distribution of cells reactive with chicken anti-mouse immunoglobulin serum. *Mol. Immunol.*, 17, 959-969.

Binz, H. and Wigzell, H. (1977). Antigen-binding, idiotypic T-lymphocyte receptors. *Contemp. Topics Immunobiol.*, 7, 113-117.

Burckhardt, J. J., Guggisberg, E., and von Fellenberg, R. (1974). Thymus immunoglobulin receptors. Ontogenic development, response to antigenic stimulation and their possible role in the regulation of "aggressive" immune reactions. *Immunology*, 26, 521-538

Cone, R. E. (1981). Molecular basis for T lymphocyte recognition of antigens. *Prog. Allergy*, in press.

Cone, R. E., Feldmann, M., Marchalonis, J. J., and Nossal, G. J. V. (1974). Cytophilic properties of surface immunoglobulin of thymus-derived lymphocytes. *Immunology*, 26, 49-60.

Crumpton, M. J., Marchalonis, J. J., Haustein, D., Atwell, J. L., and Harris, A. W. (1976). Plasma membrane of a murine T cell lymphoma: Surface labeling, membrane isolation, separation of membrane proteins and distribution of surface label amongst these proteins. *Aust. J. Exp. Biol. Med. Sci.*, 54, 303-316.

DeLuca, D., Warr, G. W., and Marchalonis, J. J. (1979). The immunoglobulin-like T-cell receptor. II. Codistribution of Fab determinants and antigen on the surface of antigen-binding lymphocytes of mouse thymus. *J. Immunogenetics*, 6, 359-372.

Hammerling, U., Pickel, H. G., Mack, C., and Masters, D. (1976). Immunochemical study of an immunoglobulin-like molecule of murine T lymphocytes. *Immunochemistry*, 13, 533-538.

Hunt, S. M. and Marchalonis, J. J. (1974). Radioiodinated lymphocyte surface glycoproteins: Concanavalin A binding proteins include surface immunoglobulin. *Biochem. Biophys. Res. Commun.*, 61, 1227-1233.

Ju, S. -T., Gray, A., and Nisonoff, A. (1977). Frequency of occurrence of idiotypes associated with anti-p-axophenylarsonate antibodies arising in mice immunologically suppressed with respect to a cross-reactive idiotype. *J. Exp. Med.*, 145, 540-556.

Marchalonis, J. J. (1975). Lymphocyte surface immunoglobulins. *Science*, 190, 20-29.

Marchalonis, J. J. (1976). Isolated-radioiodinated surface immunoglobulins of murine bone-marrow derived lymphocytes which bind the 2,4-dinitrophenyl hapten. *Immunochemistry*, 13, 667-670.

Marchalonis, J. J. (1977a). *The Lymphocyte, Parts 1 and 2.* Marcel Dekker, Inc., New York.

Marchalonis, J. J. (1977b). *Immunity in Evolution.* Harvard University Press, Cambridge, Mass.

Marchalonis, J. J., Cone, R. E., and Santer, V. (1971). Enzymic iodination. A probe for accessible surface proteins of normal and neoplastic lymphocytes. *Biochem. J.*, 124, 921-927.

Marchalonis, J. J., Cone, R. E., and Atwell, J. L. (1972). Isolation and partial characterization of lymphocyte surface immunoglobulins. *J. Exp. Med.*, 135, 956-971.

Marchalonis, J. J., Warr, G. W., Smith, P., Begg, G. S., and Morgan, F. J. (1979). Structural and antigenic studies of an idiotype-bearing murine antibody to the arsonate hapten. *Biochemistry*, 18, 560-565.

Marchalonis, J. J., Warr, G. W., Wang, A. -C., Burns, W. H., and Burton, R. C. (1980a). An Fab-related surface component of some normal and neoplastic human and marmoset T cells. *Mol. Immuno.*, 17, 877-891.

Marchalonis, J. J., Warr, G. W., Rodwell, J. D., and Karush, F. (1980b). Surface component of primate thymus-derived lymphocytes related to a heavy chain variable region. *Proc. Natl. Acad. Sci. USA*, 77, 3625-3629.

Marchalonis, J. J., Warr, G. W., Santucci, L. A., Szenberg, A., von Fellenberg, R., and Burckhardt, J. J. (1980c). The immunoglobulin-like T cell receptor. IV. Quantitative cellular assay and partial characterization of a heavy chian cross-reactive with the Fd fragment of serum μ chain. *Mol. Immunol.*, 17, 985-999.

Marchalonis, J. J. and Wang, A. -C. (1981). A marmoset T-lymphocyte protein related to defined human serum immunoglobulin and fragments. J. Immunogenetics, in press.

Moroz, C. and Hahn, J. (1973). Cell surface immunoglobulin of human thymus cells and its *in vitro* biosynthesis. *Proc. Natl. Acad. Sci USA*, 70, 3716-3720.

Moseley, J. M., Beatty, E. A., and Marchalonis, J. J. (1979). Molecular properties of T-lymphoma immunoglobulin. II. Peptide composition of the heavy chain. *J. Immunogenetics*, 6, 1-18.

Pacifico, A. and Capra, J. D. (1980). T. cell hybrids with arsonate specificity. I. Initial characterization of antigen-specific T. cell products that bear a cross-reactive idiotype and determinants encoded by the murine major histocompatibility complex. *J. Exp. Med.*, 152, 1289-1301.

Rajewsky, K. and Eichmann, K. (1977). Antigen receptors of T helper cells. *Contemp. Topics Immunobiol.*, 7, 69-112.

Tung, A. S. and Nisonoff, A. (1975). Isolation from individual A/J mice of anti-p-azophenylarsonate antibodies bearing a cross-reactive idiotype. *J. Exp. Med.*, 141, 112-126.

Vitetta, E. S. and Uhr, J. W. (1975). Immunoglobulin receptors revisited. *Science*, 189, 964.

Warr, G. W., Marton, G., Szenberg, A., and Marchalonis, J. J. (1978). Reactions of chicken antibodies with immunoglobulins of mouse serum and T cells. *Immunochemistry*, 15, 615-622.

Warr, G. W., DeLuca, D., and Marchalonis, J. J. (1979). The immunoglobulin-like T-cell receptor. III. Binding of the arsonate hapten by idiotype-bearing T cells and antibody is blocked by avian antibody to $(Fab')_2$. *Mol. Immunol.*, 16, 735-738.

INSULIN RECEPTOR REGULATION[1]

Richard A. Galbraith and Maria G. Buse

Departments of Medicine and Biochemistry
Medical University of South Carolina
171 Ashley Avenue
Charleston, South Carolina 29425

1. This research was supported in part by the American Diabetes Association, the South Carolina State Appropriations for Biomedical Research, and Grant AM 02001 from NIAMDD.

I. HISTORICAL PERSPECTIVE

Insulin was discovered by Banting and Best in 1922 and despite intensive world-wide efforts, its mechanism of action is unknown. The first evidence that insulin's biological effects were mediated by interaction with cell membranes emerged in 1949 (Levine et al., 1950; Stadie et al., 1950) and following a dormant period, the concept was confirmed by Pastan et al. (1966), Cuatrecasas (1969), and Kono and Barham (1971a). The latter year also saw the emergence of the technique for preparing monoiodo-insulin with full biological activity (Freychet et al., 1971) which heralded the last decade's direct study of insulin's interaction with its receptor.

II. MEASUREMENT OF INSULIN RECEPTORS

Most insulin binding assays are performed on isolated suspensions of cells or cell membranes. Generally, the amount of [^{125}I] insulin added is kept constant and unlabelled insulin is added at concentrations from 0 through the physiological range (0-500μU/ml). Analogous to a radioimmunoassay, after reaction, the bound [^{125}I] insulin is separated from the free hormone usually by centrifugation of cells or filtration of membranes. The percentage of radioactivity bound is termed total binding. With each assay, a nonspecific binding point is run and the percentage of nonspecific binding subtracted from total binding to yield specific binding. The amount of unlabelled insulin used for the determination of nonspecific binding varies with investigator and with tissue, but is usually 10-100ng/ml. The dilution of the [^{125}I] insulin by unlabelled insulin is so great that the possibility of a labelled molecule binding to a receptor is infinitesimally small. Thus, any tissue-associated radioactivity represents adherence to sites other than receptors and is deemed nonspecific or nonsaturable binding. Plotting specific insulin bound against total insulin present yields a sigmoidal competitive insulin binding curve.

Most investigators then perform Scatchard's (1949) analysis to derive estimates of the number and affinity of insulin receptors. The data are plotted with bound insulin (B) on the abscissa against the ratio of bound to free insulin (B/F) on the ordinate. Receptor number per cell is calculated by multiplying the abscissa intercept, R_o, by Avogadro's number (6.03×10^{23}) and dividing by the cell concentration per liter. Unfortunately, Scatchard's analyses of insulin binding data are almost invariably curvilinear with an upward concavity, hence no simple slope is calculable. The significance of this observation is undecided. Two main hypotheses have been propounded. First, that the curvilinearity represents two different binding sites, one with high affinity and low capacity, and the other with low affinity and high capacity. Second, that site-site interactions between receptors dictate that with increasing occupancy, unoccupied receptors undergo a decrease in affinity. DeMeyts et al. (1973) showed that cells allowed to reach steady state binding conditions and then "infinitely diluted" to preclude rebinding, dissociated bound insulin faster in buffer containing unlabelled insulin than in buffer alone. This phenomenon has been termed negative cooperativity and a mathematical analysis has been developed to estimate affinity based on this model (DeMeyts and Roth, 1975). While this model is widely applied and is useful as a means of comparing data, it is probably an oversimplification or a more complex situation. Gliemann (1976) has demonstrated evidence for two sites in rat adipocytes, as have Olefsky and Chang (1978). Furthermore, Pollet et al. (1977) concluded that the accelerated dissociation of insulin was not dependent on fractional occupancy but ligand-ligand exchange. The precise explanation of the curvilinearity awaits further study but is likely to involve both multiple sites and cooperativity.

III. INSULIN RECEPTOR ISOLATION AND STRUCTURE

Cuatrecasas (1972) was the first to report purification of liver membrane insulin receptors by solubilization in Triton X-100 and extraction with insulin-agarose affinity chromatography. This procedure resulted in an approximately 250,000-fold purification. The technique was modified by the addition of intermediate DEAE-cellulose chromatography and the purified rat liver insulin receptor assigned a Stokes radius of 72Å, an isoelectric point of 4, and a molecular weight of 135,000 (Jacobs et al., 1977). Partial purification of the turkey erythrocyte insulin receptor was reported in 1978, with a Stokes radius of 70Å and molecular weight of 300,000. The solubilized receptor had similar binding characteristics to native membrane-bound receptor including preservation of apparent negative cooperativity (Ginsberg et al., 1978). Human anti-receptor antibodies isolated from rare patients with acanthosis nigricans and severe carbohydrate intolerance (Flier et al., 1977; Kahn et al., 1976; Jarrett et al., 1976; Flier et al., 1976, 1977; Kahn, 1977; LeMarchand-Brustel, 1978a) further established the specificity of the solubilized receptor, and as a corollary, have also been used in receptor purification by immunoprecipitation (Harrison et al., 1979). Evidence for the subunit structure of the insulin receptor was presented following sodium dodecyl sulphate polyacrylamide gel electrophoresis of [^{125}I] insulin-aryl-azide photolabelled liver receptors (Jacobs et al., 1979) or of [^{125}I] insulin disuccinimidyl suberate cross-linked fat cell receptors (Pilch and Czech, 1979, 1980a) with and without dithiothreitol. Similarly, using electrophoresis of anti-receptor antibody immunoprecipitated receptor, Lang et al. (1980), reported four sub-units, three of which were quantitatively reduced by prior exposure of the membrane to insulin.

The last year has seen a remarkable convergence of all groups working in this field. The consensus is that three forms of disulphide-linked receptors exist in adipocyte and hepatocyte membranes, with molecular weights of 350,000, 320,000, and 290,000. They contain subunit structures of $(\alpha\beta)_2$, $(\alpha\beta)$ $(\alpha\beta_1)$, and $(\alpha\beta_1)_2$, respectively. Dithiothreitol (1mM) causes reduction to 210,000 $(\alpha\beta)$ and 160,000 $(\alpha\beta_1)$ structures, while higher concentrations (50mM) result in free α, β, and β_1 fragments (Massague and Czech, 1980; Proceedings of First International Symposium on Insulin Receptors, Rome 1980). Furthermore, the $(\alpha\beta)$ and $(\alpha\beta_1)$ structures can be isolated from native liver membranes (Massague and Czech, 1980); dithiothreitol treatment of human placental membranes causes a reduction in receptor affinity (Jacobs and Cuatrecasas, 1980), and insulin binding alters the conformation of the adipocyte receptor (Pilch and Czech, 1980b). These findings suggest that there may indeed be a structural basis for negative cooperativity, but definitive proof will probably require sequencing of the subunits.

IV. INSULIN AND INSULIN RECEPTOR INTERNALIZATION AND DEGRADATION

Since the classic experiments by Cuatrecasas (1969) which showed that insulin bound to a solid phase still produced an insulin effect, there has been controversy over the role of insulin internalization. The latter has been demonstrated by electron microscope radioautography in 1M9 lymphocytes (Carpentier et al., 1978) and hepatocytes (Gorden et al., 1976; Goldfine and Smith, 1976). Goldfine et al. (1977, 1978) reported insulin internalization and suggested that not only is insulin internalized but also that it binds to the nuclear membrane.

There has been much recent interest in the mode of insulin, or insulin receptor, internalization. Pharmacological inhibition of transglutaminase activity correlates strongly with inhibition of ligand internalization (Davies et al., 1980). It has also been shown that many agents (hormones, lectins, and immunoglobulins) stimulate membrane-bound methyltransferase, resulting in accumulation of phosphatidylcholine in the outer portion of the membrane (Hirata and Axelrod, 1980). It is postulated that resultant changes in microviscosity are induced local to the receptor which facilitate lateral movement in the plane of the membrane which (in the case of cAMP-linked ligands) promotes receptor coupling to adenylate cyclase. The insulin receptor, while not coupled to adenylate cyclase, has been shown to display capping on lymphocytes (Schlessinger et al., 1980), presumably by similar lateral membrane movement.

The role of insulin subsequent to internalization is purely speculative. Hammons and Jarrett (1979) showed that inhibition of lysosomal degradation of insulin by chloroquine produced no effect on glucose transport and LeCam et al. (1979) showed that inhibition of insulin internalization by methylamine in rat hepatocytes had no discernible effect on insulin-stimulated AIB transport. Thus available evidence tends to support the concept that insulin internalization is not essential for action. However, there is little doubt that internalization is related to insulin degradation. Insulin degradation may occur at the membrane by release of non-specific proteases (Olefsky et al., 1975) or by receptor-mediated degradation of bound insulin (Terris and Steiner, 1975; Dial et al., 1977; Gliemann and Sonne, 1978). The latter pathway is quantitatively more significant and has been measured by gel filtration of solubilized cell material (Olefsky et al., 1979). That such degradation should occur in lysosomes is very reasonable and is supported by radioautographic evidence and by the observation that lysosomotrophic agents such as chloroquine decrease receptor-mediated insulin degradation (Terris et al., 1979; Hammons and Jarrett, 1980).

Another possible consequence of insulin receptor internalization is that receptors may be processed and recycled back to the membrane. Membrane recycling has been shown in cultured rat embryonic fibroblasts (Schneider et al., 1979) and this process has been suggested by Terris (1979) to apply to the insulin receptor. The latter showed that cells exposed to insulin, washed and subsequently rebound, showed no alteration in insulin receptor number by Scatchard's analysis. In parallel, she also demonstrated insulin internalization. Since cycloheximide had no effect on receptor number, she argued that receptor recycling must be occurring. However, this observation requires caution in interpretation as cycloheximide inhibits protein degradation in addition to synthesis (Pariza et al., 1976).

V. INSULIN RECEPTOR REGULATION

Much of the data concerning receptor regulation has been the result of insulin binding studies in man. The inconvenience of adipose or muscle tissue biopsy was obviated by the demonstration of specific insulin receptors on circulating lymphocytes (Gavin et al., 1972). However, since it is the monocyte (Schwartz et al., 1975) which displays the greatest binding, significant yields require large quantities of blood and replacement of erythrocytes, especially if serial samples are required. The recent demonstration that erythrocytes bind insulin (Gambhir et al., 1977) has made studies easier to perform and more acceptable to patients.

In the last decade, numerous authors have reported alterations of insulin binding in an ever increasing array of diverse physiological and pathological states (for example, see Table 1). The plasma insulin concentration has been most convincingly shown to display an inverse relationship to insulin binding (Olefsky,

Table 1. Clinical insulin receptor abnormalities.

Clinical Entity	Affinity	Number
Obesity		Decrease
Anorexia Nervosa		Increase
Juvenile Diabetes		Increase
Maturity Diabetes		Decrease
Cushing's Syndrome	Decrease	
Growth Hormone Deficiency	Decrease	
Acromegaly	Increase	
Acanthosis Nigrigans + Insulin Receptor Antibodies: Type A		Decrease
Type B	Decrease	
Dystrophia Myotonica	Decrease	
Ageing of Erythrocytes		Decrease

1976a; Barr et al., 1976; Vigneri et al., 1978; Roth, 1976; Freychet, 1976). In humans, decreased insulin binding (down-regulation) has been shown within 5 hr of euglycemic hyper-insulinemia by Insel et al. (1980) in erythrocytes. Down-regulation has also been demonstrated *in vitro* in cultured cells (Gavin et al., 1974) and *in vivo* in experimental animals with hyperinsulinemia induced by repeated insulin injections (Kobayashi and Olefsky, 1978). Down-regulation is inhibited by cycloheximide (Gavin et al., 1974) and by blockage of ATP production and by disruption of microtubules (Kosmokos and Roth, 1976). Glyburide, an oral sulphonylurea hypoglycemic drug, has been shown to increase insulin binding and to partially inhibit down-regulation (Prince and Olefsky, 1980). More recently, Krupp and Lane (1981) demonstrated that insulin-induced down-regulation in chick hepatocytes causes no alteration in total extractable cellular insulin receptors, but decreases the proportion of the receptors on the cell membrane. This observation would appear to lend further evidence to the concept of receptor recycling.

The precise relationship between insulin binding and response has been elusive. Olefsky (1976b) and LeMarchand-Brustel et al. (1978) correlated insulin binding and acute biological responses in rodent adipocytes and soleus muscle. However, maximal response of glucose uptake in adipocytes requires receptor occupany of only about 10% (Kono and Barham, 1971b; Olefsky, 1978). Data suggest that at high concentrations of insulin, tissues from normal and obese animals respond identically, but at low concentrations, obese animals require higher concentrations than normals to obtain the same response via "reserve" or "spare" receptors, i.e., there is a rightward shift in the dose response curve, or decreased insulin sensitivity. With prolonged or higher levels of hyperinsulinemia, a post-receptor block can also occur which results in decreased insulin responsiveness. Evidence has been presented by Kolterman et al. (1980) that in a group of diabetic or obese patients, there is a spectrum of receptor and post-receptor blocks. Further study is obviously required but it can be concluded that decreased insulin receptor number does result in a right-shift of dose-response which can modify cellular responses to insulin. Thus, the study of factors which alter receptor number and/or affinity can very reasonably be expected to further our understanding of insulin's actions and the pathophysiology of abnormal carbohydrate tolerance.

VI. EBL STUDIES

In our laboratory, we have been studying insulin receptors in the erythroblastic leukemic (EBL) cell. EBL cells are chemically induced erythroblastic tumor cells (Huggins et al., 1970) which are passaged in rat pups where they grow in the hepatic sinusoids (Wise, 1974). They have been used as models of normal erythroblasts to study maturation of membrane function (Wise, 1976; Hempling and Wise, 1975; Bank et al., 1978) and in culture can synthesize hemoglobin under the influence of dimethylsulphoxide (Kluge et al., 1976).

We demonstrated that EBL cells have insulin receptors which are coupled to biological responses (Galbraith et al., 1980a). When compared to mature, nonnucleated erythrocytes, EBL cells display many more insulin receptors (23,000 vs. 60) with little differences in affinity. Insulin (100 μU/ml) caused an increase in both α-amino-isobutyric acid transport and uridine incorporation into RNA in EBL cells. In contrast, no response of erythrocytes to physiological doses of insulin has yet been reported. These data support the hypothesis that erythrocyte insulin receptors may be vestigial, and that receptor-response coupling is lost during maturation and loss of the nucleus.

Consistent with other studies, we also showed that insulin exerts a down-regulatory effect on EBL cells in primary culture

(Galbraith et al., 1980b). Insulin (100 μU/ml) within 1 hr at 37°C induced a 35% decrease in subsequent insulin binding; the effect was consistent with a decrease in receptor number. Additionally, we observed that the concentration of amino acids influenced insulin binding. This work was prompted by the demonstration that amino acids, notably glutamine, modified lysosomal degradation in hepatocytes (Schworer and Mortimore, 1979). In view of the postulated recycling of the receptor, and the potential role of the lysosome we explored the effect of different concentrations of amino acids on EBL cell insulin binding. Supplementation of the essential amino acids (plus glutamine) of Eagles Media with a mix of all amino acids found in rat plasma produced a 50% increase in binding after 1 hr at 37°C. Interestingly, such supplementation in the presence of insulin (100 μU/ml) produced less up-regulation than with amino acids alone and less down-regulation than with insulin alone; the effect was abolished by cycloheximide (1 μg/ml) (Galbraith, et al., 1980b).

Further study of this effect of amino acids has revealed that the increased binding is due entirely to the presence of serine in the supplemented media. Serine-mediated up-regulation is detectable within 15 min at 37°C, is abolished by cycloheximide, and by Scatchard's analysis, induces a 60% increase in receptor number with no change in affinity (Galbraith and Buse, unpubl.). In view of the abolition by cycloheximide, we also tested the effect of serine on protein synthesis. As with insulin binding, stimulation was detectable after 15 min of incubation. Comparison of serine dose response curves of insulin binding and protein synthesis showed a dissociation. Effects were first seen at 5 μM serine for both parameters but maximal stimulation was seen at 22μM for binding and 88 μM for protein synthesis, after which doses, both curves plateaued.

Several explanations of this effect of serine can be considered First, there may be defective synthesis of serine in EBL cells. Serine-requiring human marrow and leukemic cells have been reported (Regan et al., 1969) and the basis appears to be decreased activity of glycerate-3-phosphate dehydrogenase (Pizer and Regan, 1972). Second, cells may be depleted of serine during the preparatory washes and further, following inoculation of low densities in large volumes of medium (Lockhart and Eagle, 1959; Eagle, 1959). Third, serine may be compartmentalized within the cell such that extracellular serine is handled differently from intracellular serine. Any of the above mechanisms might account for decreased protein synthesis secondary to decreased amino acylation of seryl-tRNA, but none explain the dose-response dissociation and the preferential expression of the insulin receptor.

Such dissociation could be an example of post-transcriptional selective control (actinomycin D did not reduce either response). Alternatively, the serine-stimulated protein synthesis may result in synthesis of a regulatory protein(s) which accelerates insulin receptor processing or insertion into the plasma membrane. The latter hypothesis seems more credible, especially as the effect of serine is detectable within 15 min. Furthermore, the observation that cycloheximide had no effect on insulin receptor number in control 1-hr incubation tends to support the concept of receptor recycling.

Addition of serine to EBL cell incubations also caused a 30% increase in transferrin receptor number with no effect on affinity. Koumans and Daughaday (1963) reported that the stimulatory effect of somatomedin on [^{35}S] sulphate incorporation into chondroitin required serine for its expression; similarly, Plet and Berwaldnetter (1980) reported increased somatomedin stimulation of [^{3}H] thymidine incorporation into fibroblast DNA in the presence of serine. It will be of interest to determine if serine acts by a common mechanism to increase the number of various membrane receptors, perhaps through an intermediate which regulates receptor recycling.

ACKNOWLEDGEMENTS

The authors are grateful to Dr. Curtis Wise for providing the EBL cell, Ms. Sandra Tucker for expert technical assistance, and Mrs. Dot Smith and Mrs. Barbara Whitlock for secretarial help. Richard A. Galbraith is a recipient of a Juvenile Diabetes Foundation Research Fellowship.

VII. REFERENCES

Bank, H. L., Wise, W. C., Hargrove, K., and Spicer, S. S. (1978). Intramembranous particles in erythrocyte, reticulocyte, and erythroblastic leukemic cells of the rat: a model system for erythrocyte maturation. *Exp. Hemat.*, 6, 528-538.

Banting, F. G. and Best, C. H. (1922). The internal secretion of the pancreas. *J. Lab. Clin. Med.*, 7, 251-266.

Barr, S., Gordon, P., Roth, J., Kahn, C., and DeMeyts, P. (1976). Fluctuations in the affinity and concentration of insulin receptors on circulating monocytes of obese patients: Effects of starvation, refeeding, and dieting. *J. Clin. Invest.*, 58, 1123-1135.

Carpentier, J. L., Gordon, P., Amhar, M., VanObberghen, E., Kahn, C. R., and Orci, L. (1978). ^{125}I-insulin binding to cultured human lymphocytes: initial localization and fate of hormone determined by quantitative electron microscope autoradiography. *J. Clin. Invest.*, 6, 1057-1070.

Cuatrecasas, P. (1969). Interaction of insulin with the cell membrane: The primary action of insulin. *Proc. Natl. Acad. Sci. (USA)*, 63, 450-457.

Cuatrecasas, P. (1971). Properties of the insulin receptor of isolated fat cell membranes. *J. Biol. Chem.*, 246, 7625-7374.

Cuatrecasas, P. (1972). Affinity chromatography and purification of the insulin receptor of liver cell membranes. *Proc. Natl. Acad. Sci. (USA)*, 69, 1277-1281.

Davies, P. J. A., Davies, D. R., Levitzki, A., Maxfield, F. R., Milhaud, P., Willingham, M. C., and Pastan, I. H. (1980). Transglutaminase is essential in receptor mediated endocytosis of α_2 macroglobulin and polypeptide hormones. *Nature*, 283, 162-167.

DeMeyts, P., Roth, J., Neville, D. M. Jr., Gavin, J. R. III, and Lesniak, M. A. (1973). Insulin interactions with its receptors. Experimental evidence for negative cooperativity. *Biochem. Biophys. Res. Commun.*, 55, 154-161.

DeMeyts, P. and Roth, Jr. (1975). Cooperativity in ligand binding: a new graphic analysis. *Biochem. Biphys. Res. Commun.*, 66, 1118-1126.

Dial, L. K., Miyamoto, S., and Arquilla, E. R. (1977). Modulation of ^{125}I-insulin degradation by receptors in liver plasma membranes. *Biochem. Biophys. Res. Commun.*, 74, 545-552.

Eagle, H. (1959). Amino acid metabolism in mammalian cell cultures. *Science*, 130, 432-437.

Flier, J. S., Kahn, C. R., Roth, J., and Bar, R. S. (1975). Antibodies that impair insulin receptor binding in an unusual diabetic syndrome with severe insulin resistance. *Science*, 190, 63-65.

Flier, J. S., Kahn, C. R., Jarrett, D. B., and Roth, Jr. (1976). Characterization of antibodies to the insulin receptor. A cause of insulin-resistant diabetes in man. *J. Clin. Invest.*, 58, 1442.

Flier, J. S., Kahn, C. R., Jarrett, D. B., and Roth, J. (1977). Autoantibodies to the insulin receptor. Effect on the insulin receptor interaction in IM-9 lymphocytes. *J. Clin. Invest.* 60, 748.

Freychet, P., Roth, J, and Neville, D. M. Jr. (1971). Monoiodoinsulin: Demonstration of its biological activity and binding to fat cells and liver membranes. *Biochem. Biophys. Res. Commun.*, 43, 400-408.

Freychet, P. (1976). Interactions of polypeptide hormones with cell membrane specific receptors: studies with insulin and glucagon. *Diabetologia*, 12, 83-100.

Galbraith, R. A., Wise, W. C., and Buse, M. G. (1980a). Insulin binding and response in erythroblastic leukemic cells. *Diabetes*, 29, 571-578.

Galbraith, R. A., Tucker, S., Wise, W. C., and Buse, M. G. (1980b)

Insulin binding to erythroblastic leukemic cells is decreased by insulin and increased by amino acids. *Biochem. Biophys. Res. Commun.*, 96, 1434-1440.

Gambhir, K., Archer, J., and Carter, L. (1977). Insulin radioreceptor assay for human erythrocytes. *Clin. Chem.*, 23, 1590-1595.

Gavin, J. R. III, Archer, J. A., Lesniak, M. A., Gordon, P., and Roth, J. (1972). Hormone-receptor interactions in circulating cells. *J. Clin. Invest.*, 51, 35a.

Gavin, J. R. III, Roth, J1, Neville, D. M. Jr., DeMeyts, P., and Buell, D. N. (1974). Insulin dependent regulation of insulin receptor concentrations. A direct demonstration in cell culture. *Proc. Natl. Acad. Sci. (USA)*, 71, 84-88.

Ginsberg, B. H., Cohen, R. M., Kahn, C. R., and Roth, J. (1978). Properties and partial purification of the detergent-solubilizaed insulin receptor. A demonstration of negative cooperativity in micellar solution. *Biochim. Biophys. Acta*, 542, 88-100.

Gliemann, J. (1976). Two groups of insulin receptors in rat adipocytes, one of them related to insulin degradation. *Diabetologia*, 12, 393 (Abstract).

Gliemann, J. and Sonne, O. (1978). Binding and receptor-mediated degradation of insulin in adipocytes. *J. Biol. Chem.*, 253, 7857-7863.

Goldfine, I. D. and Mith, G. J. (1976). Binding of insulin to isolated nuclei. *Proc. Natl. Acad. Sci. (USA)*, 73, 1427-1431.

Goldfine, I. D., Smith, G. J., Wong, K. Y., and Jones, A. L. (1977). Cellular uptake and nuclear binding of insulin in human cultured lymphocytes: Evidence for potential intracellular sites of insulin action. *Proc. Natl. Acad. Sci. (USA)*, 74, 1368-1372.

Goldfine, I. D., Jones, A. L., Hradek, G. T., Wong, K. Y., and Mooney, J. A. (1978). Entry of insulin into human cultured lymphocytes: Electron microscope autoradiographic analysis. *Science*, 202, 760-762.

Gordon, P., Carpentier, J. L., Freychet, P., Cam, A. L., and Orci, L. (1978). Intracellular translocation of Iodine-125-labelled insulin: Direct demonstration in isolated hepatocytes. *Science*, 200, 782-785.

Hammons, G. T. and Jarrett, L. (1979). Abstracts, *Diabetes*, 39th Annual Meeting, Los Angeles, June, p. 389.

Hammons, G. T. and Jarrett, L. (1980). Lysosomal degradation of receptor-bound ^{125}I-insulin by rat adipocytes. *Diabetes*, 29, 475-486.

Harrison, L. C., Flier, J. S., Roth, J., Karlsson, F. A., and Kahn, C. R. (1979). Immunoprecipitation of the insulin receptor: A sensitive assay for receptor antibodies and a specific technique for receptor purification. *J. Clin. Endocrinol. Metab.*, 48, 59-65.

Hempling, H. G. and Wise, W. C. (1975). Maturation of membrane function: the permeability of the rat erythroblastic leukemic cell to water and to non-electrolytes. *J. Cell Physiol.*, 85, 195-208.

Hirata, F. and Axelrod, J. (1980). Phospholipid methylation and biological signal transmission. *Science*, 209, 1082-1090.

Huggins, C. B. and Sugiyama, T. (1966). Induction of leukemia in rat by pulse doses of 7, 12 dimethylbenz(a) anthracene. *Proc. Natl. Acad. Sci. (USA)*, 55, 74-81.

Huggins, C., Grand, L., and Oka, H. (1970). Hundred day leukemia: preferential induction in rat by pulse-doses of 7, 8, 12-trimethylbenz(a) anthracene. *J. Exp. Med.*, 131, 321-330.

Insel, J. R., Kolterman, O. G., Saekow, M., and Olefsky, J. M. (1980). Short-term regulation of insulin receptor affinity in man. *Diabetes*, 29, 132-139.

Jacobs, S., Schecter, Y., Bissell, K., and Cuatrecasas, P. (1977). Purification and properties of insulin receptors from rat liver membranes. *Biochem. Biophys. Res. Commun.*, 77, 981-988.

Jacobs, S., Hazum, E., Shechter, Y., and Cuatrecasas, P. (1979). Insulin receptor: covalent labelling and identification of subunits. *Proc. Natl. Acad. Sci. (USA)*, 76, 4918-4921.

Jacobs, S. and Cuatrecasas, P. (1980). Disulphide reduction converts the insulin receptor of human placenta to a low affinity form. *J. Clin. Invest.*, 66, 1424-1427.

Jarrett, D. B., Roth, J., Kahn, C. R., and Flier, J. S. (1976). Direct method for detection and characterization of cell surface receptors for insulin by means of ^{125}I-labelled autoantibodies against the insulin receptor. *Proc. Natl. Acad. Sci. (USA)*, 73, 4115.

Kahn, C. R., Flier, J. S., Bar, R. S., Archer, J. A., Gordon, P., Martin, M. M., and Roth, J. (1976). The syndromes of insulin resistance and acanthosis nigricans. Insulin receptor disorders in man. *N. Engl. J. Med.*, 294, 739-745.

Kahn, C. R., Baird, K., Flier, J. S., and Jarrett, D. B. (1977). Effects of autoantibodies to the insulin receptor on isolated adipocytes. Studies of insulin binding and insulin action. *J. Clin. Invest.*, 60, 1094-1106.

Kluge, N., Ostertag, W., Sugiyama, T., Arndt-Jovin, D., Steinheider, G., Furusawa, M., and Dube, S. K. (1976). Dimethylsulphoxide-induced differentiation and hemoglobin synthesis in tissue cultures of rat erythroleukemia cells transformed by 7, 12-dimethylbenz(a) anthracen. *Proc. Natl. Acad. Sci. (USA)*, 73, 1237-1240.

Kobayshi, M. and Olefsky, J. M. (1978). Effect of experimental hyperinsulinemia on insulin binding and glucose transport in isolated rat adipocytes. *Am. J. Physiol.*, 235, 932.

Kolterman, O. G., Insel, J., Saekow, M., and Olefsky, J. M. (1980). Mechanisms of insulin resistance in human obesity. Evidence for receptor and post-receptor defects. *J. Clin. Invest.*, 65, 1272-1284.

Kono, T. and Barham, F. W. (1971a). Insulin-like effects of trypsin on fat cells. Localization of the metabolic steps and cellular site affected by the enzyme. *J. Biol. Chem.*, 246, 6204-6209.

Kono, T. and Barham, F. W. (1971b). The relationship between the insulin binding capacity of fat cells and the cellular response to insulin. *J. Biol. Chem.*, 246, 6210-6216.

Kosmokos, F. C. and Roth, J. (1976). Cellular basis of insulin-induced loss of insulin receptors. Abstracts of the 58th Annual Meeting of the Endocrine Society, June, p. 69.

Koumans, J. and Daughaday, W. H. (1963). Amino acid requirements for the partially purified sulfation factor. *Trans. Assoc. Amer. Physicians*, 76, 152-162.

Krupp, M. and Lane, M. D. (1981). On the mechanism of ligand-induced down-regulation of insulin receptor level in the liver cell. *J. Biol. Chem.*, 256, 1689-1694.

Lang, U., Kahn, C. R., and Harrison, L. C. (1980). Subunit structure of the insulin receptor of the human lymphocyte. *Biochemistry*, 19, 64-70.

LeCam, A., Maxfield, F., Willingham, M. C., and Pastan, I. H. (1979). Insulin stimulation of amino acid transport in isolated rat hepatocytes is independent of hormone internalization. *Biochem. Biphys. Res. Commun.*, 88, 873-881.

LeMarchand-Brustel, Y. P., Gordon, P., Flier, J. S., Kahn, C. R., and Freychet, P. (1978a). Anti-insulin receptor antibodies inhibit insulin binding and stimulate glucose metabolism in skeletal muscle. *Diabetologia*, 14, 311.

LeMarchand-Brustek, Y., Jean Renaud, B, and Freychet, P. (1978b). Insulin binding and effects in isolated soleus muscle in lean and obese mice. *Am. J. Physiol.*, 234E, 348.

Levine, R., Goldstein, M. S., Huddleston, B., and Klein, B. (1950). Action of insulin on the 'permeability' of cells to free hexoses, as studied by its effects on the distribution of galactose. *Am. J. Physiol.*, 163, 70-76.

Lockhart, R. Z. and Eagle, H. (1959). Requirements for growth of single human cells. *Science*, 129, 252-254.

Massague, J. and Czech, M. P. (1980). Multiple redox forms of the insulin receptor in native liver membranes. *Diabetes*, 29, 945-947.

Olefsky, J. M. (1975). Effect of dexamethasone on insulin binding, glucose transport and glucose oxidation of isolated rat adipocytes. *J. Clin. Invest.*, 56, 1499-1508.

Olefsky, J. M., Johnson, J., Lin, F., Edwards, P., and Barr, S. (1975). Comparison of ^{125}I insulin binding and degradation to isolated rat hepatocytes and liver membranes. *Diabetes*, 27, 801-810.

Olefsky, J. (1976a). Decreased insulin binding to adipocytes and circulating monocytes from obese subjects. *J. Clin. Invest.*, 57, 1165-1172.

Olefsky, J. M. (1976b). The effects of spontaneous obesity on insulin binding, glucose transport and glucose oxidation of isolated rat adiopocytes. *J. Clin. Invest.*, 57, 842-851.

Olefsky, J. M. and Chang, H. (1978). Insulin binding to adipoçytes. Evidence for functionally distinct receptors. *Diabetes*, 27, 946-958.

Olefsky, J. M., Kobayashi, M., and Chang, H. (1979). Interactions between insulin and receptors following the initial binding event: Functional heterogeneity and relationships to insulin degradation. *Diabetes*, 28, 460-470.

Pastan, I., Roth, J., and Macchio, V. (1966). Binding of hormone to tissue: The first step in polypeptide hormone action. *Proc. Natl. Acad. Sci. (USA)*, 56, 1802-1809.

Pariza, M. W., Butcher, F. R., Kletzien, R. F., Becker, J. E. and Potter, V. R. (1976). Induction and decay of glucagon-induced amino acid transport in primary cultures of adult rat liver cells: paradoxical effects of cycloheximide and puromycin. *Proc. Natl. Acad. Sci. (USA)*, 73, 4511-4515.

Pilch, P. F. and Czech, M. P. (1979). Interaction of cross-linking agents with the insulin effector system of isolated fat cells. *J. Biol. Chem.*, 254, 3375-3381.

Pilch, P. F. and Czech, M. P. (1980a). The subunit structure of the high affinity insulin receptor. *J. Biol. Chem.*, 255, 1722-1731.

Pilch, P. F. and Czech, M. P. (1980b). Hormone binding alters the conformation of the insulin receptor. *Science*, 210, 1152-1153.

Pizer, L. I. and Regan, J. D. (1977). Basis for the serine requirement in leukemic and normal human leukocytes. Reduced levels of the enzymes in the phosphorylated pathway. *J. Natl. Cancer Inst.*, 48, 1897-1900.

Plet, A. and Berwald-netter, Y. (1980). The relative contribution of somatomedin to the serum-stimulated growth of human fibroblasts. *Biochem. Biophys. Res. Commun.*, 94, 744-754.

Pollet, R. J., Standaert, M. L., and Haase, B. R. (1977). Insulin binding to the human lymphocyte receptor: Evaluation of the negative cooperative model. *J. Biol. Chem.*, 252, 5828-5834.

Prince, M. J. and Olefsky, J. M. (1980). Direct *in vitro* effect of a sulphonylurea to increase human fibroblast insulin receptors. *J. Clin. Invest.*, 66, 608-611.

Regan, J. D., Vodopick, H., and S. Takeda. (1969). Serine requirement in leukemic and normal blood cells. *Science*, 163, 1452-1453.

Roth, J., Kahn, C. R., Lesniak, M. A., Gordon, P., DeMeyts, P., Megysei, K., Neville, D. M., Gavin, J. R. III, Soll, A. H., Freychet, I. D., Bar, R. S., and Archer, J. A. (1976). Receptors for insulin, NSILA-S, and growth hormone:

applications to disease states in man. *Recent Prog. Horm. Res.*, 31, 95-139.
Scatchard, G. (1949). The attractions of proteins for small molecules and ions. *Ann. N.Y. Sci. (USA)*, 51, 660-672.
Schlessinger, J., VanObberghen, E., and Kahn, C. R. (1980). Insulin and antibodies against insulin receptor cap on the membrane of cultured human lymphocytes. *Nature*, 286, 729-731.
Schneider, Y. J., Tulkens, P., deDuve, C., and Trouet, A. J. (1979). Fate of plasma membrane during endocytosis. *J. Cell. Biol.*, 82, 449-474.
Schwartz, R. H., Bianco, A. R., Handwerger, B. S., and Kahn, C. R. (1975). Demonstration that monocytes rather than lymphocytes are the insulin-binding cells in preparations of human peripheral blood mononuclear leukocytes: implications for studies of insulin resistant states in man. *Proc. Natl. Acad. Sci. (USA)*, 72, No. 2, 474-478.
Schworer, C. M. and Mortimore, G. E. (1979). Glucagon-induced autophagy and proteolysis in rat liver: mediation by selective deprivation of intracellular amino acids. *Proc. Natl. Acad. Sci. (USA)*, 76, 3169-3172.
Stadie, W. C., Haugaard, N., March, J. B., and Hills, A. G. (1950). The chemical combination of insulin with muscle (diaphragm) of normal rat. *Am. J. Med. Sci.*, 218, 265-274.
Terris, S. and Steiner, D. F. (1975). Binding and degradation of ^{125}I-insulin by rat hepatocytes. *J. Biol. Chem.*, 250, 8389-8398.
Terris, S., Hoffmann, C., and Steiner, D. F. (1979). Mode of uptake and degradation of ^{125}I-insulin by isolated hepatocytes and H4 hepatoma cells. *Can. J. Biochem.*, 57, 459-468.
Terris, S. (1979). ADA Research Meeting, Nashville, Tennessee, Oct.
Vigneri, R., Pliam, N. B., Cohen, D. C., Pezzino, V., Wong, K. Y., and Goldfine, I. D. (1978). *In vivo* regulation of cell surface and intracellular insulin binding sites by insulin. *J. Biol. Chem.*, 253, 8192-8197.
Wise, W. C. (1974). A transplantable erythroblastic stem-cell leukemia. *J. Natl. Cancer Instit.*, 52, 611-612.
Wise, W. C. (1976). Maturation of membrane function: transport of amino acid by rat erythroid cells. *J. Cell. Physiol.*, 87, 199-211.

EVOLUTION OF RECEPTORS[1]

Thomas C. Cheng

Marine Biomedical Research Program and
Department of Anatomy (Cell Biology)
Medical University of South Carolina
P.O. Box 12559 (Fort Johnson)
Charleston, South Carolina 29412

---

[1]The original information included herein has resulted from research supported by grants (PCM-8020884, PCM-8208016) from the National Science Foundation.

I. INTRODUCTION

In considering the evolution of cell surface receptors, it would appear essential to recall that the functional attributes of these molecular configurations involve more than recognition of self from nonself, i.e., as related to immunity. It has long been recognized independently that such receptors play important roles in fertilization, embryonic development, activity of the nervous system, and regulation of growth and development by hormones. More recently, the role of surface receptors in the mode of action of toxins and pharmacologic agents has been documented.

Prior to delving into some thoughts relative to the evolution of surface receptors, it needs to be pointed out that for obvious reasons, the greatest advances in our understanding of cellular recognition have been made with vertebrate cells. In several systems, particularly neurotransmitters, hormones, and antigens, it has been possible to demonstrate directly the existence of receptors associated with cell surfaces. Almostly concurrently, additional contributions have revealed more acceptable models of the structure and organization of cell membranes and possible mechanisms by which exogenous signals can be transduced across the cell membrane to induce or regulate the cell's responses. All of these advances, of course, must be considered as important continuations of our understanding of gene action. In other words, once the genetic code was deciphered, one of the major thrusts in modern biology has been to understand how the genetic program of a specialized cell becomes triggered and how it eventually becomes expressed. In pursuing this line of "common denominator" research, it became obvious that inroads had to be made into understanding the molecular basis of recognition between cells or between cells and exogenous materials, biotic and abiotic.

Although Metchnikoff originally proposed his theory of cellular immunity based on studies on marine invertebrates, the greatest advances in modern immunology, in which cell surface recognition plays a major role, have been made in vertebrate systems. Nevertheless, there appears to be an acceleration in interest in invertebrate internal defense mechanisms. This interest is due in part to the necessity of understanding how medically and economically important invertebrates confront parasites and other foreign materials, and in part because of interest in the phylogeny of immune mechanisms. It is my intent to review some salient aspects of cell surface recognition among a few major groups of invertebrates and the protochordates and to offer some thoughts relative to phylogeny of surface receptors.

As a result of numerous classical studies, there can be no doubt that invertebrates are capable of distinguishing between self and nonself. This, of course, implies that the chemical receptors on the surfaces of invertebrate cells, primarily hemocytes, are capable of recognition. It needs to be emphasized, however, that the steric molecular basis of invertebrate recognition probably is not one analogous to antigen-antibody or enzyme-substrate binding. It thus follows that a variety of operative mechanisms could be present but with similar end results, including modulations in ionic concentration and associated differences in electrical potentials and differences in concentration of the signal. An example of the latter has been contributed by Gerisch and Malchow (1975) who have provided evidence suggesting that the morphogenesis of the unicellular form of *Dictyostelium* into a plasmodium occurs when a signal consisting of chemotactic pulses of cAMP is received. Furthermore, they reported that the receptor system detects changes of cAMP concentration with time rather than concentration *per se*. Thus, in a way, this phenomenon resembles cholinergic synaptic transmission in the nervous system.

To illustrate the idea that although surface receptors exist on invertebrate cells, these may have quite different physicochemical bases, reviewed below are some known features of surface receptors among the protozoa, sponges, molluscs, insects, enchinoderms, and protochordates.

II. PROTOZOA

Historically, Reynolds (1924) was the first to demonstrate that a protozoan, *Arcella polypora*, can recognize self. Specifically, he reported that pseudopodial fragments experimentally severed from one amoeba will re-fuse with the parent amoeba. However, if cytoplasmic fragments of different amoebae are permitted to make contact with an amoeba, fusion usually does not occur. In other words, the amoeba is capable of recognizing self and this results in fusion. This historically interesting study implies the occurrence of recognition sites on amoebae or it could mean the exchange of cytoplasmic information, which will be mentioned again later in connection with sponge cells.

More recent evidence by Korn and Weisman (1967) suggests that protozoan phagocytosis, at least the initial stimulus, is physical rather than chemical. Specifically, these investigators have reported that *Acanthamoeba* will internalize latex beads after contact is made between bead and cell membrane. The only variable found was the sizes of beads, thus suggesting that the basis of recognition is physical. This, however, does not appear to be the case among all protozoa. For example, Hirshon (1969) has found that *Paramecium bursaria* discriminates among different types of particles of similar size when presented in equal concentrations. This suggests that there is a chemical basis to recognition.

The available information relative to protozoan surface receptors is yet too scanty to permit any definitive statements relative to their nature and how they function. Nevertheless, recent studies involving the demonstration of lectin-binding sites, especially on parasitic species (Seed et al., 1976; Dwyer, 1974, 1976, 1977; Dwyer and D'Alesandro, 1976; Trissl et al., 1977; Kelly et al., 1976; Takahashi and Sherman, 1980; and others) have revealed that there are carbohydrate-containing membrane sites. These, as in other types of cells that have been studied, are capable of mobility. However, the functional attributes of these lectin-binding sites remain essentially unknown other than by inference.

While considering surface recognition, the pioneering work of the late Tracy Sonneborn on mating types of paramecia should at least be mentioned. Irrespective of the signal, the ability of these protozoans to discriminate between compatible and incompatible mating types suggests the presence of rather specific surface receptors.

III. PORIFERA

Since sponge cells are favored models by certain developmental biologists interested in cell-cell recognition and adhesion, more is known about receptors on these cells. The topic has been expertly reviewed by Burger et al. (1978) and Steinberg (1978) and need not be reviewed in detail here.

It is of interest to point out that Burger et al. (1975) have suggested that two basically different mechanisms are involved in cell-cell recognition: (1) cell surface recognition via recognizing molecules, and (2) recognition via the exchange of intracellular information between the two cells (Fig. 1). It is recognized that the first is the one commonly considered while the latter, if true, remains to be explored in depth.

Relative to surface recognizing molecules, Barondes and Rosen (1976) have advanced the idea that new sugar-specific lectin-like macromolecules appear on the cell surfaces and these serve the recognition function. Evidence for this, in this author's opinion, are not convincing. On the other hand, reports on the occurrence of an aggregation factor in the case of sponge cells is interesting. In brief, commencing with Humphreys (1963), it has been demonstrated that cells of *Microciona* spp. dissociated into Ca^{++}-and Mg^{++}-free sea water release a high molecular weight factor without which the cells will no longer reaggregate. This factor has been shown to be a proteoglycan complex with a molecular weight of 21×10^6 (Cauldwell et al., 1973; Henkart et al., 1973). Humphreys (1963) and Mascona (1968) have studied such a factor from several

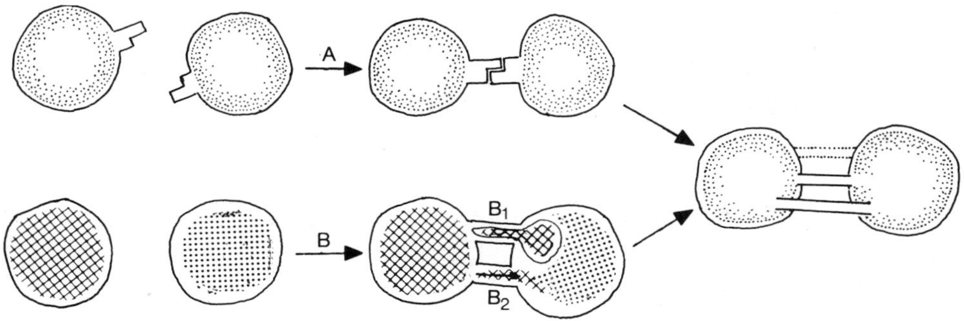

Fig. 1. Two possible mechanisms for recognition. A. This sequence dipicts recognition occurring via the interaction of specific surface molecules. B. This sequence depicts an alternative mechanism whereby information signals are passed between cells by vesicular exchange (B_1) or via gap or other junctions (B_2). The scheme depicted in B_2 is likely to have a short life span and can be considered as mutual probing sometimes seen at the ruffling edges of moving cells. Both types of specific interactions, i.e., A and B, can lead either to secondary, stabilizing linkages, i.e., intercytoplasmic, intermembranous, or extracellular linkages as depicted by the cell doublet to the right. (Modified after Burger et al., 1975).

species of sponges and have found them to be species-specific, i.e., they promote the reaggregation of homospecific cells only. This, however, has been shown by Sara et al. (1966) and Turner and Burger (1973) not to be universally true. In some instances, specificity apparently is a matter of concentration of the aggregation factor. For example, Turner and Burger (1973) have reported if the factor from *Haliclona occulata* is present at high concentrations, it can cause the aggregation of *Microciona prolifera* cells. These two sponges belong to two separate orders.

The second component of cell aggregation among sponge cells is the baseplate (Fig. 2). This is the term assigned to a component of the cell surface that recognizes the aggregation factor. In other words, it is the surface receptor. Burger and Jumblatt (1977) have demonstrated that the baseplate isolated from *Microciona prolifera* is heat sensitive (60°C for 10 min) and nondialyzable. It is stable towards EDTA, to a pH range of 3-12, and to freezing, thawing, and lyophilization. Burger and Jumblatt have suggested that it is a peripheral rather than an integral protein since it

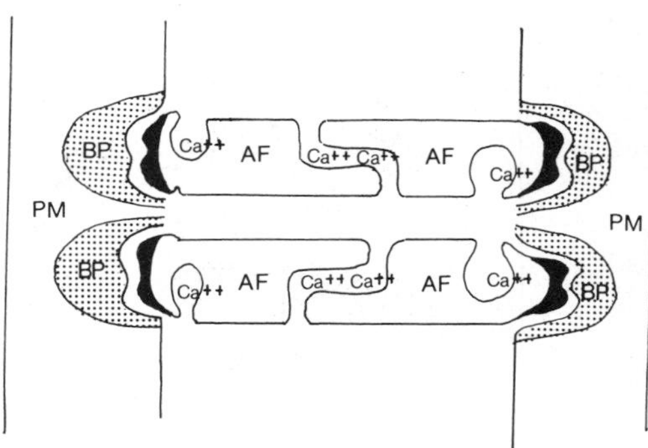

Fig. 2. A model for aggregation factor (AF)-mediated cell recognition in *Microciona prolifera* molecules held in stable association by a Ca^{++} ions containing carbohydrate moieties (black areas) that interact with surface baseplates (BP). PM = plasma membranes. (Modified after Burger et al., 1975).

can be removed from the cell surface by mild, nonpeptonic treatment and also because it cannot be sedimented at 105,000g for 90 min.

That the baseplate indeed is the receptor for the aggregation factor in sponge cell clumping has been clearly demonstrated by Weinbaum and Burger (1973) who coupled isolated baseplate to Sepharose beads and found that the beads are efficiently aggregated by the aggregation factor as were intact cells.

It is apparent from the brief review presented that surface receptors in the form of a protein baseplate occurs on sponge cells, at least certain species.

IV. MOLLUSCA

Although the occurrence of surface recognition is a well established phenomenon on molluscan hemocytes, a great deal needs to be learned about the molecular basis of the receptor sites. That such exists is implied by the numerous studies that have demonstrated that nonself materials, especially parasites, experimentally or naturally introduced into molluscs are readily phagocytosed or encapsulated. Although this is the general rule, there are exceptions. As examples, Bang (1961) reported that when hemocytes of the American oyster, *Crassostrea virginica*, are exposed to marine bacteria, *Vibrio* spp., some are phagocytosed while others are not; Michelson (1963) reported the absence of cellular response in certain freshwater pulmonates to microsporidans; and Mackin (1951) reported that invasion of *C. virginica* by the fungus *Perkinsus marinus* and the protozoan *Nematopsis* does not result in phagocytosis. These observations suggest that the receptors on molluscan cells are capable of discriminating between self and non-self. Those recognized as self are not phagocytosed. In view of recent advances in our understanding of shared antigens between host and parasite and/or adsorption of host molecules onto the surfaces of parasites (see Brown, 1976, for review), it is not surprising that certain parasites, including pathogenic species, are recognized as self and are not phagocytosed. All this, of course, implies the existence of receptors on molluscan phagocytes.

Probes into the molecular structure of molluscan receptors has only begun. Renwrantz and Cheng (1977a,b) have started to map the topography of molluscan granulocytes by establishing the presence of carbohydrate receptors for various lectins. As a result, it is now known that sugar receptors occur on the surface of *Helix pomatia* hemocytes (Renwrantz and Cheng, 1977a), and by studying lectin-mediated attachment of erythrocytes onto the surfaces of these cells have been able to confirm this. Similarly, Schoenberg and Cheng (1980, 1981), by employing microhemadsorption assays, have demonstrated the occurrence of specific sugars on the surfaces of hemocytes of *Biomphalaria glabrata* and *Bulinus truncatus*. Also, they have demonstrated that these are quantitative differences in cell surface oligosaccharides capable of binding lectins on different strains of the freshwater pulmonate *Biomphalaria glabrata* (see Schoenberg and Cheng, 1980). These findings were not especially unexpected; nevertheless, they are among the first to reveal sugar-containing, lectin-binding sites on molluscan cells.

As everyone is aware, another way of demonstrating surface receptors on cells is to employ assays for chemotaxis. This is based on the premise that there are cell surface receptors that are capable of recognizing specific chemotactic agents. In this regard, Cheng and Howland (1979) have demonstrated that hemocytes of the oyster *C. virginica* are chemotactically attracted to certain live Gram-positive (*Bacillus megaterium* and *Micrococcus varians*) as well as Gram-negative (*Escherichia coli*) bacteria but not to others of both groups. Furthermore, there is no attraction to any heat-killed bacteria. This implies discrimination. Subsequently, Howland and Cheng (1981) have reported that the chemoattractant is associated with the cell wall of Gram-positive (*Bacillus megaterium*) and the cell envelope of Gram negative bacteria (*Escherichia coli*). It is protein with a molecular weight of approximately 10,000 which is devoid of either a carbohydrate or a lipid moeity. This, of course, implies that the recognition site on the surface of oyster hemocytes is not a lectin-like molecule which has an affinity for sugars. In collaboration with Dr. Keith

Howland, we have begun to study the nature of the oyster hemocyte chemoreceptor. As shown in Fig. 3, treatment of the molluscan hemocytes with pronase at higher concentrations (0.5, 0.0, 0.5%) results in a decrease of signal recognition, thus indicating that a protein is involved. Treatment with trypsin at all five of the concentrations tested (Fig. 4) also results in a decline in

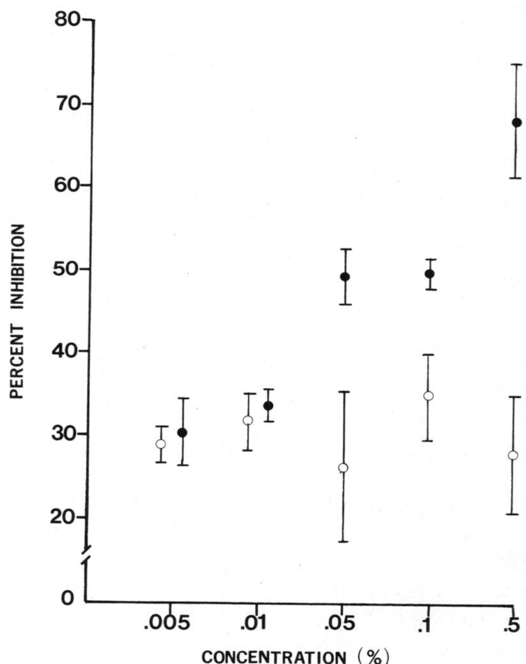

Fig. 3. Percent inhibition of random migration (o) toward sterile saline and chemotactic migration (•) toward live *Bacillus megaterium* by *Crassostrea virginica* hemocytes pretreated with five concentrations of pronase. Verticle lines represent 1 standard deviation. The experimental protocol is that presented by Cheng and Howland (1979).

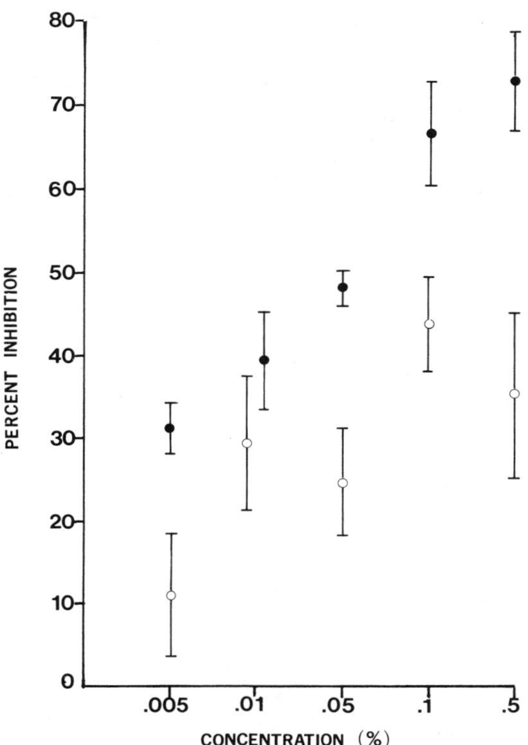

Fig. 4. Percent inhibition of random migration (o) toward sterile saline and chemotactic migration (•) toward live *Bacillus megaterium* by *Crassostrea virginica* hemocytes pretreated with five concentrations of trypsin. Verticle lines represent 1 standard deviation. The experimental protocol is that presented by Cheng and Howland (1979).

recognition, thus indicating that lysine or arginine is the amino acid in position 1. On the other hand, although treatment with chymotrypsin (Fig. 5) also resulted in significant decrease in recognition, the greatest percentages occurred at the lower concentrations (0.005, 0.01, and 0.05%). Treatment with pepsin only resulted in decreases in recognition at the highest concentrations. Other than the preliminary data presented, nothing is presently known about this molecule.

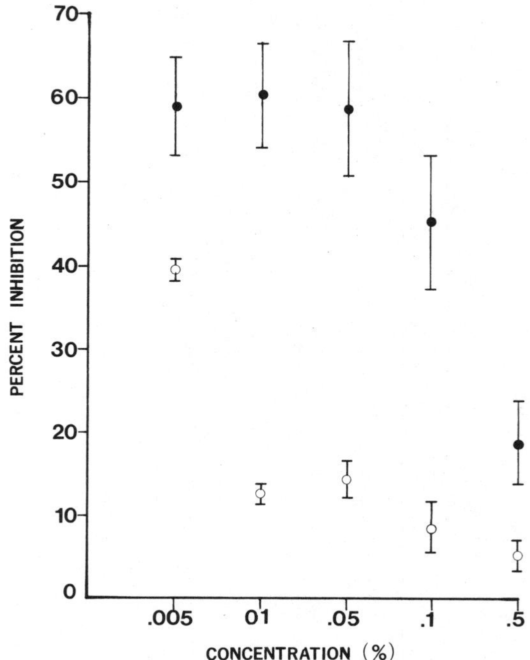

Fig. 5. Percent inhibition of random migration (o) toward sterile saline and chemotactic migration (•) toward live *Bacillus megaterium* by *Crassostrea virginica* pretreated with five concentrations of chymotrypsin. Verticle lines represent 1 standard deviation. The experimental protocol is that presented by Cheng and Howland (1979).

V. INSECTA

Although considerable attention has been directed toward understanding the internal defense mechanisms of arthropods, especially insects (see Ratcliffe and Rowley, 1979 for review). Surprisingly little work has been done on the identification of hemocyte surface receptors. Perhaps this is because such leaders in the field of insect immunology as Jones (1956) and Salt (1970) have stated dogmatically that chemotaxis between arthropod hemocytes and foreign materials does not occur; thus implying the absence of cell surface recognition sites. The above not withstanding, others have suggested the occurrence of chemotaxis involving insect hemocytes; however, many of these are inferred from the study of histopaghologic sections (Nappi, 1974; Nappi and Stoffolano, 1972). There are some direct evidences for the occurrence of insect hemocyte chemotaxis to "wound" or "injury factors" released by damaged cells (Wigglesworth, 1937; Harvey and Williams, 1961; Lea and Gilbert, 1961; Clark and Harvey, 1965; Wyatt and Linzen, 1965; Cherbas, 1973). Also, positive chemotaxis toward certain bacteria have now also been documented (Gagen and Ratcliffe, 1976; Ratcliffe and Gagen, 1976). These studies on chemotaxis are being mentioned because the occurrence of the phenomenon implies the presence of recognition sites on insect hemocytes. The nature and function of such sites, however, have not been thoroughly investigated.

Another line of research suggesting the presence of cell surface receptors in insects is that aimed at understanding phagocytosis. The following is a brief review of the most significant studies in this area from the standpoint of identification surface receptors.

The best evidence for insect hemocyte surface receptors are those contributed by Anderson (1976a,b) and Scott (1971a). As a result of studying hemocytes of *Spodoptera eridania* and *Estigmene acrea*, Anderson has shown that serum-independent attachment of hemocytes is probably mediated in part by interaction of cell surface receptors. These, however, are nonspecific. Scott (1971a) demonstrated that the attachment of *Periplanata americana* hemocytes is mainly dependent on surface receptors which are trypsin-labile.

Rabinovitch and De Stefano (1970) have reported an important observation, that is, *Galleria mellonella* hemocytes phagocytosed a greater variety of modified erythrocytes *in vivo* than *in vitro*. If surface receptors exist, as the results of Anderson (1976a,b) and Scott (1971a) suggest, then the results of Rabinovitch and De Stefano could be interpreted to mean that additional recognition factor(s) on functional *in vivo* augment the surface receptors.

Another approach that has been taken to provide evidence of the presence of insect cell surface receptors has been to employ enzymes. In this vein, Scott (1971b) has found that pretreatment of the hemocytes of the American cockroach, *Periplanata americana*, with trypsin almost completely inhibits the adherence of sheep erythrocytes without reducing the uptake of neutral red, thus indicating the insect cells were viable. Rosette formation is not restored after treating hemocytes with undiluted serum, thus suggesting that trypsin altered surface receptors for an opsonic humoral component. Scott has also shown that pretreatment with lipase causes slight enhancement of sheep erythrocyte attachment. Further evidence for the existence of cell surface receptors was provided by passively treating roach hemocytes with rabbit anti-sheep erythrocyte serum. This resulted in a marked increase in adherent sheep erythrocytes. This phenomenon did not occur when sheep erythrocytes, presensitized with antiserum, were exposed to roach hemocytes. Thus, cockroach hemocytes apparently lack a receptor site for cytophilic and noncytophilic components of mammalian immune serum. Nevertheless, Scott's results indicate surface receptors on cockroach hemocytes.

Finally, it is of interest to point out that numerous investigators (Poinar et al., 1968; Crossley, 1975; Rizki and Rizki, 1976; Ratcliffe and Gagen, 1976, 1977; Schmit and Ratcliffe, 1977) have reported that encapsulation reactions in insects is usually initiated by the release of a substance from certain types of hemocytes. Ratcliffe (1975) has now demonstrated that binding of certain foreign materials *in vitro* to certain hemocytes of *Pieris brassicae* is the result of adherence to a "sticky" layer of acid mucopolysacchride surrounding the cells. This is the material secreted by the cells. An essentially identical material that facilitates binding has been reported by Ratcliffe and Gagan (1976, 1977) and Schmit and Ratcliffe (1977). Thus, it would appear that the binding of insect granulocytes to nonself materials, at least in certain instances, is mediated by a secreted, surfacial stratum of acid muncopolysaccharide. This, as is readily recognized, is a departure from the usual surface receptors.

VI. ECHINODERMATA

Since the echinoderms, being deuterostomates, are in the direct evolutionary line leading to the vertebrates, an understanding of the nature of their surface recognition sites is of primary interest if the intent is to elucidate the evolutionary trend.

Although the types of coelomocytes of echinoderms have been studied in great detail by several investigators (Ohuye, 1934; Johnson, 1969a,b; and others), essentially nothing is known about the nature of their surface receptors. Nevertheless, it is known that at least certain of these cells are capable of recognizing self from nonself and discriminating between categories of foreign materials. For example, Johnson (1969a,b) has demonstrated that the phagocytic coelomocytes, which she designated as phagocytic leukocytes, readily phagocytose Gram-positive bacteria but not Gram-negative species. The latter are disposed of by lysis or killing. Even then, the degree and rapidity of the reaction on the part of coelomocytes vary.

Of particular interest is the fact that Johnson (1969a,b) has reported that one of the types of coelomocytes, known as vibratile cells, is rich in acid mucopholysaccharides, and under certain stress conditions, these molecules are discharged into the coelomic fluid. This is reminiscent of the secretion of an acid mucosaccharide by insect granulocytes. Also, a second type of coelomocytes, the red spherule cells, secrete the pigment echinochrome (a red napthoquinone pigment) when challenged with bacteria; and as a result, motile species of bacteria become immobilized.

VII. PROTOCHORDATA

Except for some yet unpublished studies from our laboratory on *Molgula*, as far as I have been able to determine, no studies have yet been performed to identify receptor sites on tunicate cells. It is known, however, that these primitive chordates possess phagocytic cells. For example, Freeman (1970) and Smith (1970) have both reported phagocytosis by certain cells, which Smith has designated as hyaline amoebocytes. Interestingly, Anderson (1971) has reported that when glass fragments are inserted into the branchial sac of *Molgula manhattensis*, they are encapsulated by another type of hemocyte, vanadocytes, and vanadocyte-produced tunicin. However, injected carmine and trypan blue are phagocytosed by amoebocytes. Thus it would appear that phagocytosis and encapsulation are performed by different types of cells. This suggests different types of cell surface receptors.

Taking another approach, Wright (1974) has found that the serum of *Ciona intestinalis* includes a natural agglutinin which acts on a variety of vertebrate erythrocytes *in vitro*. *In vivo*, this agglutinin facilitates clearance of experimentally introduced duck and human erythrocytes. Cooper (1976) has hypothesized that this agglutinin, plus that found by Fuke and Sugai (1972) in two solitary ascidians, *Styela plicata* and *Halocynthia hilgendorfi*, may play a role in cell-cell adherence, although experimental data are still wanting.

VIII. AN ANALYSIS

From the brief review presented, it would appear that cell surface receptors most probably occur in all invertebrates, including the protozoa and the protostomate and deuterostomate groups. Furthermore, the Porifera, which is an abbarent group, also have cell surface receptors.

Based on very limited chemical data, it would appear at this time that invertebrate and tunicate receptors, at least certain ones, may be chemically different, although the functional end results are similar. Briefly, among protozoa at least four types of receptors occur: (1) an undeciphered type of surface receptor apparently selects on a physical basis, i.e., size differences; (2) the common sugar-lectin type; (3) another unexplained type of surface receptor operative in the selection of mating types, and (4) the possible existence of cytoplasm-associated information exchange.

Among sponge cells, the proteoglycan aggregation factor-peripheral protein baseplate mechanism is wide spread, although the exchange of intracellular information, which has yet to be convincingly demonstrated, may also occur. Finally, the lectin-sugar type of receptor complex is also known to occur.

Among molluscan cells, in addition to the sugar-lectin type of binding, a protein-protein recognition system associated with chemotaxis also occurs.

Among insects cells, nonspecific surface receptors of unknown nature, "wound factors", enzyme-labile receptors, secreted acid mucopholysaccharides, as well as sugar-lectin binding are the categories of surface receptors identified thus far.

Along the deuterostomate line, echinoderm cells also display the secreted acid mucopolysaccharide type of receptor, and the sugar-lectin type of binding probably also occurs. Among protochordate cells, indirect evidences suggest the occurrence of different types of surface receptors, including the sugar-lectin binding type.

Thus, even at this primitive stage of our knowledge, it would appear that the most common type of cell surface receptor is of the lectin-sugar binding type. It exists throughout the Animal Kingdom. It is of interest to note that the Fc, Fab, and C3 receptors common to vertebrates apparently do not occur in invertebrates. In addition, other yet not fully understood types occur in each of the phyla examined. These may be unique to each group and serve highly specific functions.

IX. REFERENCES

Anderson, R. S. 1971. Cellular responses to foreign bodies in the tunicate *Molgula manhattensis* (DeKay). *Biol. Bull.*, 141, 91-98.

Anderson, R. S. 1976a. Expression of receptors by insect macrophages. In "Phylogeny of Thymus and Bone Marrow-Bursa Cells." (R. K. Wright and E. L. Cooper, eds.). pp. 27-34. Elsevier/North-Holland, Amsterdam.

Anderson, R. S. 1976b. Macrophage function in insects. *Proc. First Intl. Colloq. Invert. Pathol.*, pp. 215-219. *Queens Univ. Press, Kingston, Ontario, Canada.*

Bang, F. B. 1961. Reaction ot injury in the oyster (*Crassostrea virginica*). *Biol. Bull.*, 121, 57-68.

Barondes, S. H. and Rosen, S. D. 1976. Cellular recognition in slime molds. Evidence for its mediation by cell surface species-specific lectins and complimentary oligosaccharides. In "Surface Membrane Receptors." (R. P. Bradshaw, W. A.

Frazier, R. C. Merrell, D. I. Gottlieb. and R. A. Hogue-Angeletti, eds.) pp. 39-55. *Plenum, N.Y.*

Broron, K. N. 1976. Specificity in host-parasite interaction. In "Receptors and Recognition." (P. Cuatrecasas and M. F. Greaves, eds.). pp. 121-175. *John Wiley & Sons, N.Y.*

Burger, M. M. and Jublatt, J. 1977. Membrane involvement in cell-cell interactions: a two-component model system for cellular recognition that does not require live cells. In "Cell and Tissue Interactions." (J. W. Larsh and M. M. Burger, eds.). pp. 155-172. *Raven, N.Y.*

Burger, M. M., Turner, R. S., Kuhns, W. J., and Weinbaum, G. 1975. A possible model for cell-cell recognition via surface macromolecules. *Philos. Trans. Roy. Soc., London,* B 271, 379-393.

Burger, M. M., Burkart, W., Weinbaum, G., and Jumblatt, J. 1978. Cell-cell recognition: molecular aspects. Recognition and its relation to morphogenetic processes in general. In "Cell-Cell Recognition." pp. 1-23. *Cambridge Univ. Press, Cambridge, England.*

Cauldwell, C., Henkart, P., and Humphreys, T. 1973. Physical properties of sponge aggregation factor: a unique proteoglycan complex. *Biochemistry,* 12, 3051-3055.

Cheng, T. C. and Howland, K. H. 1979. Chemotactic attraction between hemocytes of the oyster, *Crassostrea virginica,* and bacteria. *J. Invert. Pathol.,* 33, 204-210.

Cherbas, L. 1973. The induction of an injury reaction in cultured haemocytes from saturniid pupae. *J. Insect Physiol.*, 19, 2011-2023.

Clark, R. M. and Harvey, W. R. 1965. Cellular membrane formation by plasmatocytes of diapausing cecropia pupae. *J. Insect Physiol.*, 11, 161-175.

Cooper, E. L. 1976. "Comparative Immunology." *Prentice-Hall, Englewood Cliffs, N.J.*

Crossley, A. C. S. 1975. The cytophysiology of insect blood. *Adv. Insect Physiol.*, 11, 117-222.

Dwyer, D. 1974. Lectin binding saccharides on a parasitic protozoan. *Science,* 184, 471-473.

Dwyer, D. 1976. Cell surface saccharides of *Trypanosoma lewisi.* II. Lectin-structure cytochemical detection of lectin-binding sites. *J. Cell Sci.*, 22, 1-19.

Dwyer, D. 1977. *Leishmania donovani*: Surface membrane carbohydrates of promastigotes. *Exp. Parasit.*, 41, 341-358.

Dwyer, D. M. and D'Alesandro, P. A. 1976. The cell surface of *Trypanosoma musculi* bloodstream forms. II. Lectin and immunologic studies. *J. Protozool.*, 23, 262-271.

Freeman, G. 1970. The reticuloendothelial system of tunicates. *J. Reticuloend. Soc.*, 7, 183-194.

Fuke, M. T. and Sugai, T. 1972. Studies on the naturally occurring hemagglutinin in the coelomic fluid of an ascidian. *Biol. Bull.*, 143, 140-149.

Gagen, S. J. and Ratcliffe, N. A. 1976. Studies in the *in vivo* cellular reactions and fate of injected bacteria in *Galleria mellonella* and *Pieris brassicae* larvae. *J. Invert. Pathol.*, 28, 17-24.

Gerisch, G. and Malchow, D. 1976. Cyclic AMP receptors and the control of cell aggregation in *Dictyostelium*. *Adv. Cyclic Nucleot. Res.*, 7, 49-68.

Harvey, W. R. and Williams, C. M. 1961. The injury metabolism of the cecropia silkworm. I. Biological amplification of the effects of localized injury. *J. Insect Physiol.*, 7, 81-99.

Henkart, P. S., Humphreys, S., and Humphreys, T. 1973. Characterization of sponge aggretation factor: a unique proteoglycan complex. *Biochemistry*, 12, 3045-3050.

Hirshon, J. B. 1969. The response of *Paramecium bursaria* to potential endocellular symbionts. *Biol. Bull.*, 136, 33-42.

Humphreys, T. 1963. Chemical and *in vitro* reconstruction of sponge cell adhesion. I. Isolation and functional demonstration of the components involved. *Dev. Biol.*, 8, 27-47.

Johnson, P. T. 1969a. The coelomic elements of sea urchins (*Strongylocentrotus*). II. Cytochemistry of the coelomocytes. *Histochemie*, 17, 213-231.

Johnson, P. T. 1969b. The coelomic elements of sea urchins (*Strongylocentrotus*). III. *In vitro* reaction to bacteria. *J. Invert. Pathol.*, 13, 42-62.

Jones, J. C. 1956. The hemocytes of *Sarcophaga bullata*. Parker *J. Morph.*, 99, 233-257.

Kelly, P., Cotman, C. W., Gentry, C., and Nicolson, G. L. 1976. Distribution and mobility of lectin receptors on synaptic membranes of identified neurons in the central nervous system. *J. Cell Biol.*, 71, 487-496.

Korn, E. D. and Weisman, R. A. 1967. Phagocytosis of latex beads by *Acanthamoeba*. II. Electron microscopic study of the initial events. *J. Cell Biol.*, 34, 219-227.

Lea, M. S. and Gilbert, L. E. 1966. The hemocytes of *Hyalophora cecropia* (*Lepidoptera*). *J. Morph.*, 118, 197-215.

Mackin, J. G. 1951. Histopathology of infection of *Crassostrea virginica* (Gemlin) by *Dermocystidium marinum* Mackin, Owen and Collier. *Bull. Mar. Sci. Gulf Caribb.*, 1, 72-87.

Mascona, A. A. 1968. Cell aggregation: properties of specific cell ligands and their role in the formation of multi-cellular systems. *Dev. Biol.*, 18, 250-277.

Michelson, E. H. 1963. *Plistophora husseyi* sp.n., a microsporidian parasite of aquatic pulmonate snails. *J. Insect Pathol.*, 5, 28-38.

Nappi, A. J. 1974. Insect hemocytes and the problems of host recognition of foreignness. *Contemp. Top. Immunobiol.*, 4, 207-224.

Nappi, A. J. and Stoffolano, J. G. Jr. 1972. Distribution of haemocytes in larvae of *Musca domestica* and *Musca autumnalis* and possible chemotoaxis during parasitization. *J. Insect Physiol.*, 18, 169-179.

Ohuye, T. 1934. On the coelomic corpuscles in the body fluid of some invertebrates. I. Reaction of the leucocytes of a holothuroid, *Caudina chilensis* (J. Muller) to vital dyes. *Hatai Sci. Rep. Tohoku Imp. Univ. 4th ser. (Biol.)* 99L, 47-59.

Poinar, G. O. Jr., Leutenegger, R., and Gotz, P. 1968. Ultrastructure of the formation of a melanotic capsule in *Diabrotica* (Coleoptera) in response to a parasitic nematode (Mermithidae). *J. Ultrastruct. Res.*, 25, 293-306.

Rabinovitch, M. and Destefano, M. J. 1970. Interactions of red cells with phagocytes of the wax moth (*Galleria mellonella*, L.) and mouse. *Exp. Cell. Res.*, 59, 272-282.

Ratcliffe, N. A. 1975. Spherule cell-test particles interactions in monolayer cultures of *Pieris brassicae* hemocytes. *J. Invert. Pathol.*, 26, 217-223.

Ratcliffe, N. A. and Gagen, S. J. 1976. Cellular defense reactions of insect hemocytes *in vivo*: Nodule formation and development in *Galleria mellonella* and *Pieris brassicae* larvae. *J. Invert. Pathol.*, 28, 373-382.

Ratcliffe, N. A. and Gagen, S. J. 1977. Studies on the *in vivo* cellular reactions of insects. An ultrastructural analysis of nodule formation in *Galleria mellonella*. *Tissue Cell*, 9, 73-85.

Ratcliffe, N. A. and Rowley, A. F. 1979. Role of hemocytes in defense against biological agents. In "Insect Hemocytes. Development, Forms, Functions, and Techniques." (A. P. Gupta, ed.). pp. 331-414. *Cambridge Univ. Press, Cambridge, England.*

Renwrantz, L. R. and Cheng, T. C. 1977a. Identification of agglutinin receptors on hemocytes of *Helix pomatia*. *J. Invert. Pathol.*, 29, 88-96.

Renwrantz, L. R. and Cheng, T. C. 1977b. Agglutinin-mediated attachment of erythrocytes to hemocytes of *Helix pomatia*. *J. Invert. Pathol.*, 29, 97-100.

Reynolds, B. D. 1924. Interactions of protoplasmic masses in relation to the study of heredity and environment in *Arcella polypora*. *Biol. Bull*, 46, 106-142.

Rizki, M. T. M. and Rizki, R. M. 1976. Cell interactions in hereditary melanotic tumor formation in *Drosophila*. *Proc. First Intl. Colloq. Invert. Pathol.* pp. 137-141. *Queens University Press, Kingston, Ontario, Canada.*

Salt, G. 1970. "The Cellular Defence Reactions of Insects." *Cambridge Uni. Press, London.*

Sara, M. Liaci, L., and Melone. N. 1966. Biospecific cell aggregation in spones. *Nature*, 210, 1167-1168.

Schmit, A. R. and Ratcliffe, N. A. 1977. The encapsulation of foreign tissue implants in *Galleria mellonella* larvae. *J. Insect Physiol.*, 23, 175-184.

Schoenberg, D. A. and Cheng, T. C. 1980. Lectin-binding specificities of hemocytes from two strains of *Biomphalaria glabrata* as determined by microhemadsorption assays. *Dev. Comp. Immunol.*, 4, 617-628.

Schoenberg, D. A. and Cheng, T. C. 1981. Lectin-binding specificities of *Bulinus truncatus* hemocytes as demonstrated by microhemadsorption. *Div. Comp. Immunol.*, 5, 145-149.

Scott, M. T. 1971a. Recognition of foreignness in invertebrates. II. *In vitro* studies of cockroach phagocytic haemocytes. *Immunology*, 21, 817-827.

Scott, M. T. 1971b. A naturally occurring hemagglutinin in the hemolymph of the American cockroach. *Arch. Zool. Exp. Gen.*, 112, 73-80.

Smith, J. J. 1970. The blood cells and tunic of the ascidian *Holocynthia aurantium* (Pallas). I. Hematology, tunic morphology, and partition of cells between blood and tunic. *Biol. Bull.*, 138, 354-378.

Seed, T. M., Seed, J. R., and Brindley, D. 1976. Surface properties of bloodstream trypanosomes (*trypanosoma brucei brucei*). *Tropenmed. Parasit.*, 27, 202-212.

Steinberg, M. S. 1978. Cell-cell recognition in multicellular assembly: levels of specificity. In "Cell-Cell Recognition." pp. 25-49. *Cambridge Univ. Press, Cambridge, England.*

Takashashi, Y. and Sherman, I. W. 1980. *Plasmodium lophurae:* Lectin-mediated agglutinin of infected red cells and cytochemical fine-structure detection of lectin binding sites on parasite and host cell membranes. *Exp. Parasit.*, 49, 233-247.

Trissl, D., Martinez-Palomo, A., Arguello, C. de la Torre, M., and de la Hoz, R. 1977. Surface properties related to concanavalin A-induced agglutination. A comparative study of several *Entamoeba* strains. *J. Exp. Med.*, 145, 652-665.

Turner, R. S. and Burger, M. M. 1973. Involvement of a carbohydrate group in the active site for surface guided reassociation. *Nature*, 244, 509-510.

Weinbaum, G. and Burger, M. M. 1973. A two-component system for surface guided reassociation of animal cells. *Nature*, 244, 510-512.

Wigglesworth, V. B. 1937. Wound healing in an insect, *Rhodnius prolixus* (Hemiptera). *J. Exp. Biol.*, 14, 364-381.

Wright, R. K. 1974. Protochordate immunity. I. Primary immune response of the tunicate *Giona intestinalis* to vertebrate erythrocytes. *J. Invert. Pathol.*, 24, 29-36.

Wyatt, G. R. and Linzen, B. 1965. The metabolism of ribonucleic acid in *Cecropia* silkmouth pupae in diapause, during development and after injury. *Biochim. Biphys. Acta*, 103, 588-600.

SURFACE ANTIGENS OF MURINE MELANOMA CELLS

Douglas M. Gersten*, Vincent J. Hearing**, and
John J. Marchalonis***

*Department of Pathology and National Biomedical Research Foundation, Georgetown University Medical Center Washington, D.C. 20007
** Dermatology Branch, National Cancer Institute, National Institutes of Health, Bethesda, Maryland 20205
*** Department of Biochemistry, Medical University of South Carolina, 171 Ashley Avenue, Charleston, South Carolina 29425

I. INTRODUCTION

A great deal of attention has recently been focused on host-tumor interactions, i.e. the host's response to bearing a tumor (see Shapot, 1979 for review). A logical consequence of this research has been the consideration of immune factors, both cellular and humoral, in the regulation of tumor growth and metastatic spread (Lewis et al., 1976). Central to the concept of tumor immunology is the fundamental consideration of the tumor cell as an antigen. It follows then, that the identification of antigenic moieties and the determination of their cellular location are of primary importance.

The tumor which we have studied is the B16 murine melanoma, syngeneic to C57Bl/6 mice. The evidence which suggests that this tumor carriers immunologic determinants is two-fold. First, C57Bl/6 mice may be immunized outright to B16 preparations (Fidler et al., 1977; Bystryn, 1978). Second, B16 cells injected into tumor bearing hosts behave differently from those administered identically to naive hosts (Fidler et al., 1977). These observations imply that B16 cells contain moieties which are unique to the tumor cells, while absent from other C57Bl/6 cells. Moreover, these molecules should be immunogenic.

II. APPROACHES TO ANTIGEN ISOLATION

The search for and identification of these molecules has been approached from two different directions in our laboratories. One approach has been the preparation of a xenoantiserum to B16 melanoma cells (Fig. 1). This antiserum, when suitably absorbed, demonstrated apparent specificity for the B16 and 1735 mouse melanomas, while having no reactivity toward any of the other cell types tested (Gersten and Marchalonis, 1979). A solid phase immune absorbant was prepared from this purified antiserum. Those proteins shed from cell culture material toward which the antiserum activity was directed were recovered by immune-affinity chromatography. Electrophoretic characterization of the affinity-purified proteins shed in culture demonstrated the presence of two major bands with apparent molecular weights of 65,000 and 50,000 (Fig. 2). A minor component of approximately 15,000 daltons was occasionally but not always demonstrable.

Electron microscope observation of the fine structure of the melanosomes of normal melanocytes indicates a regular arrangement of microfibrils, on which the melanin is deposited (Fig. 3). In both murine and human melanoma cells, the melanosomal fine structure is variable, ranging from irregular fibrils to complete absence of them. The fact that these fibrils are proteinaceous forms the basis

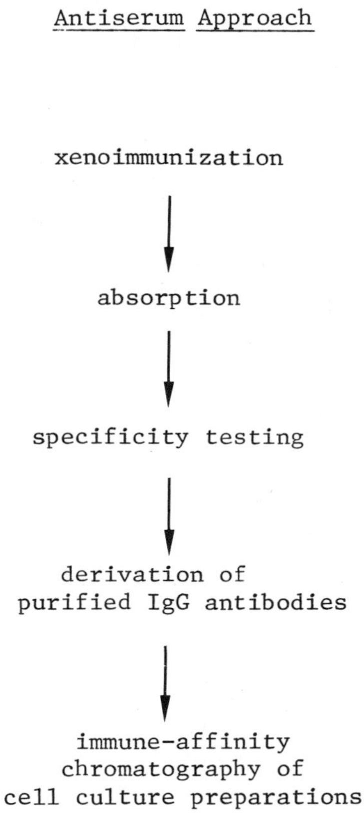

Antiserum Approach

xenoimmunization

↓

absorption

↓

specificity testing

↓

derivation of
purified IgG antibodies

↓

immune-affinity
chromatography of
cell culture preparations

Fig. 1. Antiserum approach to the isolation of B16 antigens. Goats were immunized to B16 melanoma cells, grown as a solid tumor. The antisera was absorbed with glutaraldehyde fixed EI4 lymphoma cells, and then tested for specificity against a variety of targets. On the basis of the targets used, it was determined that the remaining antibody was not directed against mouse xenoantigens, C type viruses, histocompatibility antigens, or melanin. The purified antibodies were derivatized to a solid phase matrix. Immune-affinity chromatography was used to recover the antigens from cell extracts or those shed into the cell culture supernatants.

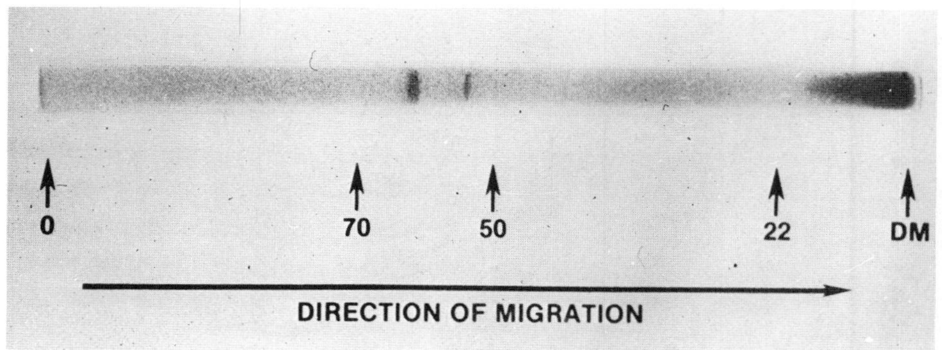

Fig. 2. SDS-polyacrylamide gel electrophoresis (Laemmli, 1970) in 10% gels of proteins purified by immune-affinity chromatography from cell culture supernatants. 0 = origin, DM = dye marker. (reprinted with permission of Academic Press).

of the second approach to antigen isolation (Fig. 4). Here, the melanosomes of normal C57Bl/6 melanocytes and B16 melanoma cells have been extracted and isolated, solubilized, and the proteins compared with each other (Klingler et al., 1976). The major protein of normal melanosomes (termed C700) and its corresponding protein in B16 melanosomes (B700) have been studied in detail by N-terminal analysis, C-terminal analysis, amino acid analysis and fingerprint analysis (Hearing and Nicholson, 1980). These studies suggested that B700 is a deletion variant of C700, possible resulting from the defective splicing of introns. It has been demonstrated that B700, but not C700, gives a positive response when tested for antigenicity in a syngeneic system (Table 1) (Kerney et al., 1977; Hearing et al., 1978).

III. IDENTITY OF 65,000 DALTON PROTEIN AND B700

The molecular weight of B700 is 68,000. Since the reduced molecular weight of the primary peak in Fig. 2 is approximately 65,000, it was important to compare the two preparations with each other. In Fig. 5, four preparations are compared by polyacrylamide gel electrophoresis under nondenaturing conditions with

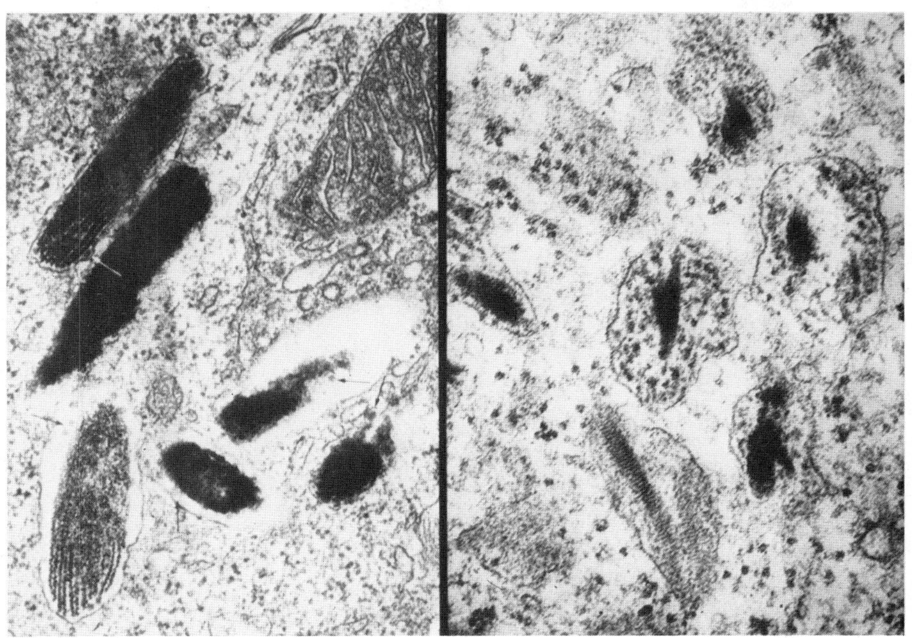

Fig. 3. Electron micrographs of melanosomes from normal C57El/6 melanocytes and B16 melanoma cells, obtained under the conditions given in Fig. 4.

the Tris: glycine buffer system 400 (Ugel et al., 1971); protein material shed into serumless medium from cultured B16 and S91 melanoma cells, 65,000 dalton protein purified from Triton X-100 extracts of cultured whole B16 cells by immune affinity chromatography and preparative gel electrophoresis, and purified B700 protein. The results demonstrate that (1) S91 Cloudman melanoma cells shed protein into serumless medium essentially the same as do B16 cells, and (2) all four preparations behave similarly in this gel system.

The same four preparations were subjected to sodium dodecyl sulfate (SDS)-polyacrylamide gel electrophoresis under fully denaturing conditions with the Laemmli (1970) buffer system and are shown in Fig. 6. Similarly, these proteins were compared with respect to their isoelectric points by isoelectric focusing

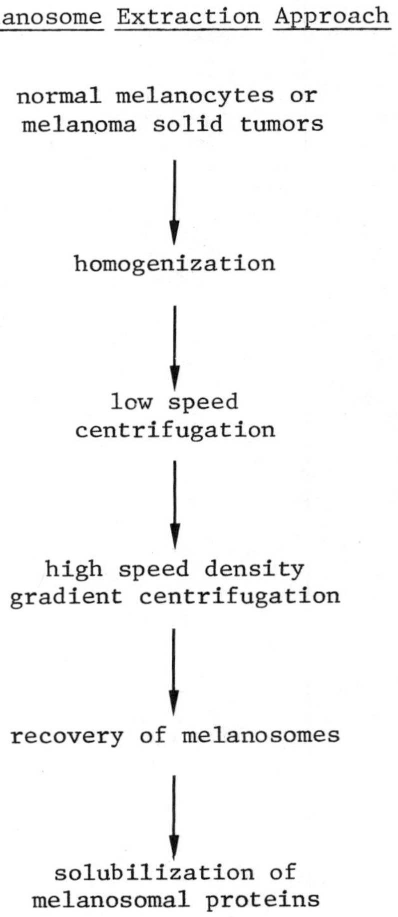

Melanosome Extraction Approach

normal melanocytes or
melanoma solid tumors

↓

homogenization

↓

low speed
centrifugation

↓

high speed density
gradient centrifugation

↓

recovery of melanosomes

↓

solubilization of
melanosomal proteins

Fig. 4. Melanosome extraction approach. The source of normal melanocytes was dorsal epidermis from 5-day-old C57B1/6 mice. Malignant melanocytes were obtained from E16 melanom tumors grown *in vivo* in the thigh muscle of C57B1/6 mice. Tissues were homogenized in isotonic phosphate buffer and the homogenates centrifuged at 500 x g for 10 min. The supernatants were centrifuged at 10,000 x g for 20 min. The pellets were recovered and washed twice by centrifugation through 30% sucrose. The resultant pellets were solubilized in 0.1% Triton X-100 and recentrifuged at 10,000 g for 20 min. The Triton X-100 isoluble material was discarded and the supernatnats filtered through a 0.45 μm Millipore filter.

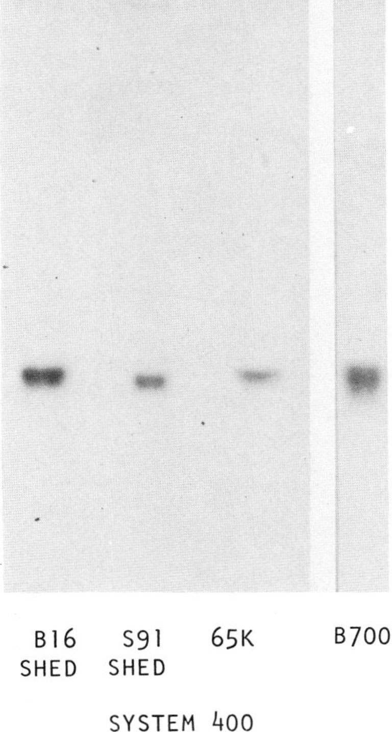

Fig. 5. Comparison of four different preparations under non-denaturing conditions in Tris-glycine buffer system 400 gels. Left to right, protein material shed into serumless medium from cultured B16 line F10 cells; shed materials from S91 mouse melanoma cultures; 65,000 dalton protein purified from Triton X-100 extracts of B16 F10 whole cells by immune affinity chromatography and preparative gel electrophoresis; purified B700 protein.

with the method of Righetti and Drysdale (1971) and are shown in Fig. 7. As in Fig. 5, the four preparations bear electrophoretic identity to each other both on the basis of their size and their charge. The similarity of electrophoretic behavior strongly suggests the identity of the 65,000 dalton preparation isolated by immune-affinity chromatography from cultured tumor cells and the B700 preparation isolated from melanosomes of B16 cells growing as a solid tumor.

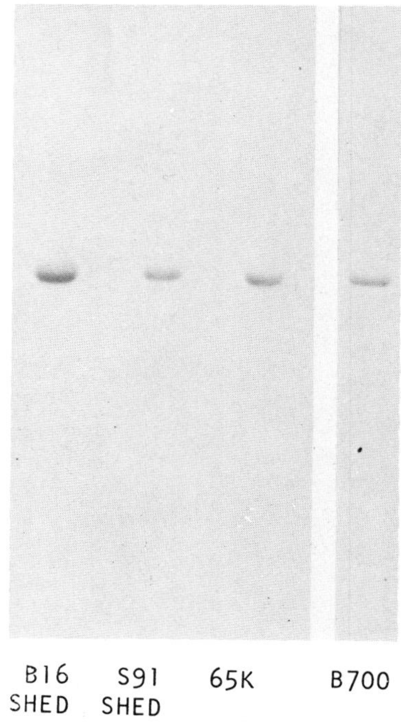

Fig. 6. Comparison of preparations as listed in Fig. 5 in the SDS polyacrylamide gel system of Laemmli (1970).

IV. CELLULAR LOCATION OF THE B16 ANTIGENS

Having established the electrophoretic identity of the 65,000 dalton and B700 preparations, we may proceed to review the evidence for the cellular location of the antigens, and the forms they assume. Two tests, leukocyte migration inhibition and lymphocyte stimulation, have been used in the past to study the location of immunologically significant B16 proteins (Kerney et al., 1977; Hearing et al., 1978). In these experiments, B16 tumors were homogenized, fractionated by density gradient ultracentrifugation, and the subcellular fractions tested for lymphocyte stimulating and migration inhibitory activity. The results were compared to those obtained from identical fractions from normal tissues (Table 1).

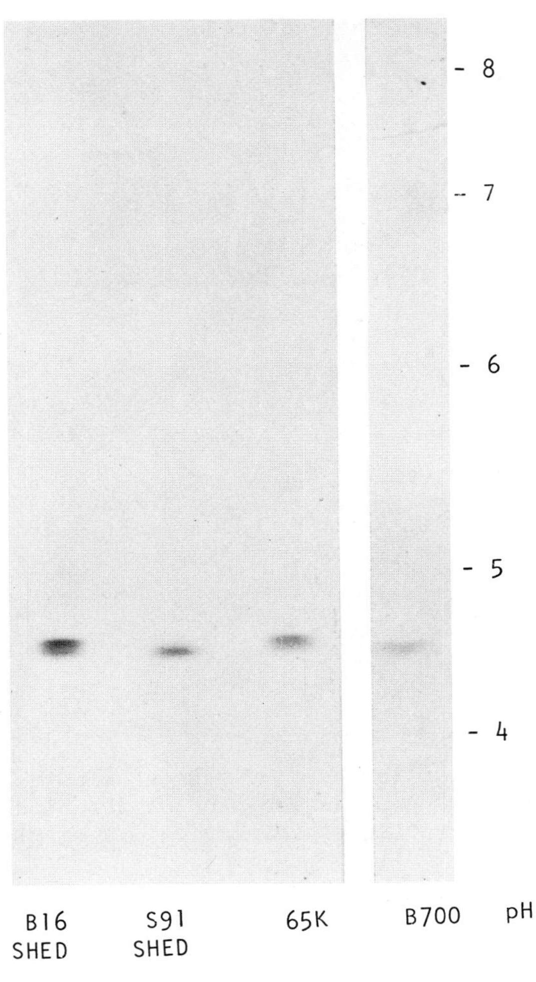

Fig. 7. Comparison of preparations as listed in Fig. 5 by isoelectric focusing.

Table 1. Cellular Location of Antigenic Material.

Fraction[1]	Migration Inhibition Assay[2]		Lymphocyte Stimulation Assay[3]	
	Normal	Melanoma	Normal	Melanoma
Crude Homogenate	0.76	0.69	0.5	1.7
Crude Melanosomes	1.06	0.73	0.2	2.3
Soluble Fraction	0.91	0.95	0.1	1.2
Microsomal Fraction	1.03	0.78	0.2	1.5
Purified Melanosomes	0.96	0.78	1.1	1.6
Autologous Muscle	0.92	0.86	0.9	1.1
B700 purified protein	1.05	0.72	0.4	2.9

[1]Subcellular fractions isolated by sucrose density gradient ultracentrifugation; B700 protein isolated as described by Hearing and Nicholson, 1980. Protein concentration 0.1 mg/ml.

[2]Migration inhibition assay carried out on macrophages isolated from normal and melanoma bearing donors and incubated with the noted subcellular fraction; index < 0.8 considered positive.

[3]Lymphocyte stimulation assay carried out lymphocytes isolated from normal and melanoma bearing donors and incubated with the noted subcellular fraction; index > 1.3 considered positive.

Several points are noteworthy from these data: activity in both tests was observed to be associated with the membranous fractions of the tissues, but not with the soluble fraction or with autologous muscle tissue. On an equal protein basis, the purified B700 protein had the highest specific activity of all the fractions. Activity was absent in the purified C700 protein from normal melanocytes.

During the course of antigen detection and purification by the xenoantiserum approach, additional information regarding the location and nature of the material was obtained. First, cytotoxicity and antibody binding studies were performed to assess the specificity of the absorbed antiserum. The materials used as targets were either viable B16 cells in suspension, or whole cells fixed with glutaraldehyde to a plastic substratum (Gersten and Marchalonis, 1979). That the antibodies bound to whole cell preparations indicates that the antigen is located at the cell surface.

Analysis of the material recovered by immune-affinity chromatography of B16 culture supernatants (Fig. 2) indicates the presence of two bands of 65,000 and 50,000 daltons with the former present in greater proportion than the latter. Other experiments (data not shown) suggest that the proportion of the two bands relative to each other varies with culture conditions. The relationship of the 65,000 and 50,000 dalton forms to each other is unclear at this time. The two possibilities which are most likely, in our estimation, are: (1) the 50,000 form is a breakdown product of the 65,000 form, or (2) the two forms may represent differential glycosylation of the same polypeptides. Although both these explanations are consistent with shared antigenic determinants on both these polypeptides, the evidence presented below favors the second possibility. In Figs. 2 and 8 it is seen that affinity purified, shed material is either in the 65,000 or 50,000 form. Affinity purified material from Triton X-100 extracts of cultured cells is present as a high molecular weight complex of over 200,000 daltons under nondenaturing conditions. This complex breaks down partially when denatured with SDS alone (Fig. 8a), and upon complete reduction with heat and mercaptoethanol further dissociates to give the 50,000 form exclusively (Fig. 8b).

In Fig. 9, proteins shed into the culture supernatant have been studied by SDS polyacrylamide gel electrophoresis. Shed materials from three cell lines, B16-F10, S91 amelanotic, and S91 melanotic mouse melanomas, have been compared. It is seen that by far the primary component of the shed materials is the 65,000 alton antigen under these culture conditions. Whether shedding to this extent occurs *in vivo* has not yet been determined.

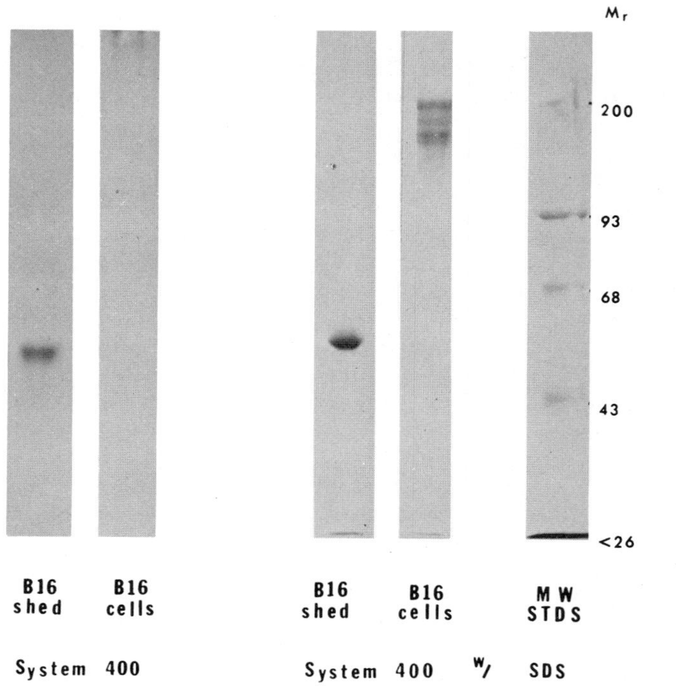

Fig. 8A
Analysis of affinity purified proteins from B16 F10 melanoma cultures. Comparison of shed proteins versus triton X-100 extracts from whole cells. (A) System 400 (non-denaturing) and System 400 with 0.1% (non-reducing) gels. (B) Laemmli (reducing) gels.

V. DISCUSSION

We have isolated antigens unique to B16 murine melanoma from material shed in culture, extracted by Triton X-100 from cultured cells, and from melanosomes of solid tumors. The major melanosomal protein of B16 tumors (B700) has a molecular weight of 68,000, is antigenic when tested in a syngeneic system, and is probably a multiple deletion mutant of its counterpart (C700) in normal melanocytes. The proteins isolated from cell culture material have approximate reduced molecular weights of 65,000 and 50,000. The 65,000 dalton form and B700 bear electrophoretic identity when studied under different conditions.

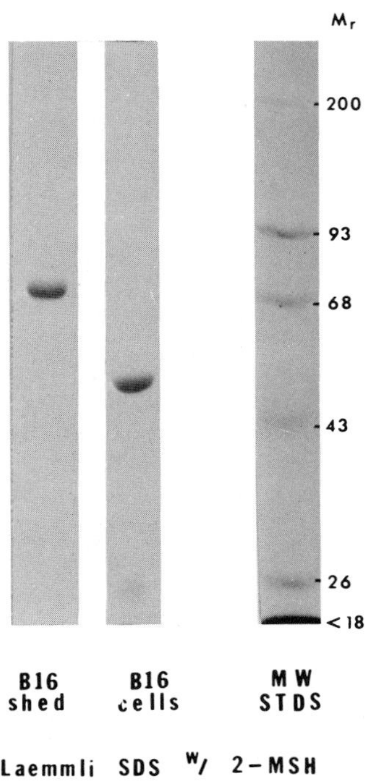

Fig. 8B

The antigens are present on all membranous structures of the cell studied, including the plasma membrane. They are conspicuously absent (or extremely low in titer) from the soluble fraction of the cell and are released readily into the culture supernatant upon incubation in serumless medium.

The relative proportions of the 65,000 and 50,000 forms which can be recovered by immune affinity chromatography, vary both with the culture conditions and the source. When shed material is the source of antigens, the 65,000 dalton form usually predominates. Furthermore, the antigens are the predominant molecular species of shed material in culture. The proportion of the 50,000 dalton form frequently but not always

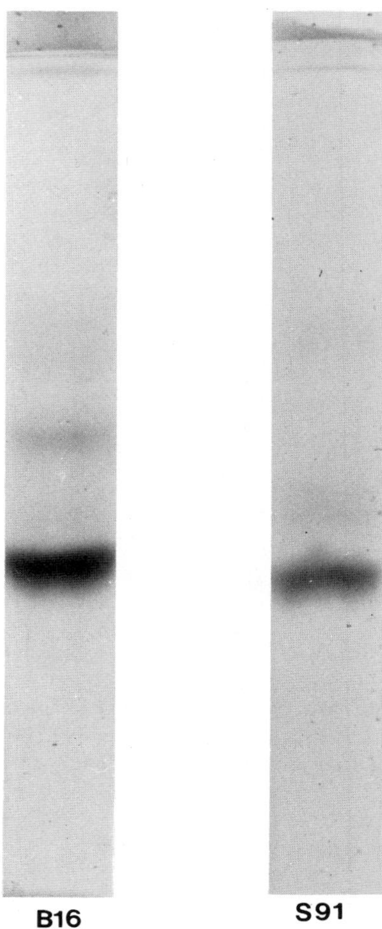

Fig. 9. SDS-polyacrylamide gel electrophoresis of unfractionated shed material from B16 and S91 culture supernatants.

increases when cell extracts rather than shed material are studied. Under nonreducing conditions, the cell extract material has an aggregate molecular weight of greater than 200,000. Similar observations have been made previously (Bystryn et al., 1974).

Reports of high molecular weight antigens (100,000 to 400,000 daltons) may be found in the literature for both human and mouse melanomas (Bystryn et al., 1974; Gupta et al., 1979; Morgan et al., 1981; Galloway et al., 1981). Other studies have reported sizes which are in the 50,000 to 70,000 dalton range (Gersten and Marchalonis, 1979; Hearing and Nicholson, 1979; Embleton et al., 1980; Khosravi et al., 1980; Bhavandan et al., 1980). Although a unifying hypothesis regarding the pathobiology of melanoma antigens is not yet possible, these conflicting reports are readily understood in light of the above data. Melanoma cells in culture appear to synthesize membrane associated antigens and release them into the extracellular medium. These are present as high molecular weight complexes inside the cell. In processing the antigens for release, the cell breaks down the complex into monomeric units of 65,000 or 50,000 daltons. If the antigens are isolated in the shed form from the medium, the lower molecular weight forms will be recovered under native or denaturing conditions.

Melanomas are highly aggressive tumors. In mice, an inoculum of relatively few B16 cells can kill the host, and only weak immunologic responses have been observed. The above data suggest a possible basis for these observations. If the antigenic proteins are capable of eliciting an immune reaction in the mouse, such a response could be readily subverted by the wholesale secretion of "decoy" antigens by the tumor cells. Two related observations support this view. First, TA3 tumor cells can escape graft rejection when injected into allogeneic hosts by a mechanism involving shedding of histocompatibility antigens (Nowotny et al., 1974). Second, blocking activity toward *in vitro* cytotoxicity tests has been ascribed to human melanoma antigens (Murray et al., 1978).

The consideration of unique cell surface molecules is important in two ways in the context of cell surface recognition in the host-tumor interaction: first, in the immune regulation of tumor growth, and second, in the fate in metastatic distribution of circulating tumor cells. Concerning the immune regulation of tumor growth, the demonstration of antigens on melanosomes, intracellular membranes, the cell surface and in the shed material is significant. It suggests that all of the following situations may contribute to the antigenic burden of the host: intact tumor cells in the primary mass or released into the cirulation, fragmented or dying cells in the necrotic center of the tumor, nondividing tumor cells shedding antigens, intact melanosomes phagocytosed by macrophages.

In considering the metastatic fate and distribution of circulating tumor cells, it has been appreciated for some time that these are dictated in part by molecules at the cell surface and *in toto* by the interaction of circulating tumor cells with host elements. The "processing" of tumor cells by the host is very rapid following their introduction into the circulation. Therefore, the temporal presentation of proteins at the cell surface and their instantaneous recognition and interaction is critical. The data presented above represent a convincing definition of unique tumor cell surface proteins, isolated by two completely unrelated approaches. This system should, therefore, prove valuable in future studies aimed at defining the specifics of host-tumor recognition and metastatic spread.

VI. REFERENCES

Bhavandan, V. P. , Kemper, J. G., and Bystryn, J. C. (1980). Purification and partial characterization of a murine melanoma associated antigen. *J. Biol. Chem.*, 255, 5745.

Bystryn, J. C. (1978). Antibody response and tumor growth in syngeneic mice immunized to partially purified B16 melanoma associated antigens. *J. Immunol.*, 120, 96.

Bystryn, J. C., Schenkein, I., Barr, S., and Uhr, J. W. (1974). Partial isolation and characterization of antigen(s) associated with murine melanoma. *J. Natl. Cancer Inst.*, 52, 1263.

Embleton, M. J., Price, M. R., and Baldwin, R. W. (1980). Demonstration and partial characterization of common melanoma associated antigen(s). *Eur. J. Cancer*, 16, 575.

Fidler, I. J., Gersten, D. M., and Riggs, C. W. (1977). Relationship of host immune status to tumor cell, arrest, distribution and survival in experimental metastasis. *Cancer*, 40, 46.

Galloway, D. R., McCabe, R. P., Pellegrino, M. A., Ferrone, S., and Reisfeld, R. A. (1981). Tumor associated antigens in spent medium of human melanoma cells: immunochemical characterization with xenoantisera. *J. Immunol.*, 125, 62.

Gersten, D. M. and Marchalonis, J. J. (1979). Demonstration and isolation of murine medlanoma associated antigenic surface proteins. *Biochem. Biphys. Res. Comm.*, 90, 1015.

Hearing, V. J. and Nicholson, J. M. (1980). Abnormal protein synthesis in malignant melanoma cells. *Cancer Eiochem. Biophys.*, 4, 59.

Hearing, V. J., Kerney, S. E., Montague, P. M., Ekel, T. M., and Nicholson, J. M. (1978). Characterization of the intracellular location of tumor associated antigens in B16 murine malignant melanoma, In "Pigment Cell, Pathophysiology of Melanocytes" (V. Riley, ed.), pp. 148-154, Karger Press, New York.

Kerney, S. E., Montague, P. M., Chretien, P. B., Nicholson, J. M., Ekel, T. M., and Hearing, V. J. (1977). Intracellular localization of tumor associated antigens in murine and human malignant melanoma. *Cancer Res.*, 37, 1519.

Khosravi, M., Preddie, E., Hartmann, D., and Lewis, M. G. (1980). Human melanoma tumor specific antigens from tumor cell plasma membranes of five melanoma patients. *Cancer Biochem. Biophys.*, 4, 195.

Klingler, W. G., Montague, P. M., and Hearing, V. J. (1976). Unique melanosomal proteins in murine melanomas, In "Pigment Cell, Melanomas, Basic Properties and Clinical Behavior" (V. Riley, ed.), pp. 1-12, Karger Press, New York.

Laemmli, U. K. (1970). Cleavage os structural proteins during the assembly of the head of bacteriophage T4. *Nature*, 227, 680.

Lewis, M. G., Hartmann, D., and Jerry, L. M. (1976). Antibodies and anti-antibodies in human malignancy: an expression of deranged immune regulation. *Ann. N.Y. Acad. Sci.*, 276, 316.

Morgan, A. C., Galloway, D. R., Imai, K., and Reisfeld, R. A. (1981). Human melanoma associated antigens. Role of carbohydrate in shedding and cell surface expression. *J. Immunol.*, 126, 365.

Murray, E., Ruygrok, S., Milton, G. W., and Hersey, P. (1978). Analysis of serum blocking factors against leukocyte dependent antibody in melanoma patients. *Int. J. Cancer*, 21, 578.

Nowotny, A., Grohsman, J., Abdelnoor, A., Rote, N., Yang, C., and Waltersdorf, R. (1974). Escape of TA3 tumors from allogeneic immune rejection: theory and experiments. *Eur. J. Immunol.*, 4, 73.

Righetti, P. G. and Drysdale, J. W. (1971). Isoelectric focusing in polyacrylamide gels. *Biochim. Biphys. Acta*, 236, 17.

Shapot, V. S. (1979). On the multiform relationships between the tumor and the host. *Adv. Cancer Res.*, 30, 89.

Ugel, A. R., Chrambach, A., and Rodbard, D. (1971). Fractionation and characterization of an oligomeric series of bovine keratohyalin by polyacrylamide gel electrophoresis. *Anal. Biochem.*, 43, 410.

MYELIN ORGANIZATION, BASIC PROTEIN STRUCTURE, AND EXPERIMENTAL DEMYELINATION

Michael J. Weise and Steven Brostoff

Department of Neurology
Medical University of South Carolina
171 Ashley Avenue
Charleston, South Carolina 29425

I. INTRODUCTION

Myelin is a membrane of central and peripheral nervous systems (CNS and PNS) which is responsible for the white color of CNS and PNS tissues. It constitutes a significant proportion of the nervous system and forms a sheath around axons which greatly enhances the efficiency and speed of nerve impulse conduction. In diseases like multiple sclerosis and the Guillian-Barre syndrome, there is destruction of this membrane and a resultant debility or death.

Reviewing the work on myelin ultrastructure and aspects of myelinogenesis (Raine, 1977) shows the following: The multilaminar structure is elaborated from the myelin forming cells (oligodendroglia in the CNS and Schwann cells in the PNS) via a flattening and expansion of processes from their plasma membrane. During the production of myelin, the cytoplasmic faces of the plasma membrane extensions abut and extrude cytoplasm while the extracellular surfaces of the extensions become apposed to form successive lamellae. These cytoplasmic and extracellular appositions are seen in electron micrographs as major dense lines (MDL) and intraperiod lines (IPL), respectively. Oligodendroglia have the capacity to myelinate the segments of several axons simultaneously while Schwann cells are dedicated to the ensheathment of a single axon segment. Both cell types have a tremendous capacity to synthesize membrane during the period of active myelination.

While the basic morphologies are similar, there are specific differences observed between the structures of CNS and PNS myelins. The periodicity of the lamellae are greater in PNS compared to CNS myelin and there are differences regarding compaction of the innermost outermost lamellae in these two membranes. Another point of difference is found in myelin structure at the nodes of Ranvier with the PNS myelin sheath showing more fine structure in the region of the node compared to that seen in the CNS counterpart.

From biochemical analyses of isolated CNS and PNS myelins (see Norton, 1977; Braun and Brostoff, 1977; Benjamins and Morell, 1978 for reviews), these membranes have been shown to be predominantly lipid (70% to 80% by weight) with significant differences in lipid composition between the membrane from CNS and PNS tissue. In these myelins, only a few types of proteins account for the majority of the total protein present. The major proteins from the CNS and PNS myelins are qualitatively different with the CNS membrane containing the myelin basic protein (MBP) and the proteolipid protein (PLP) and the PNS membrane having the P0 glycoprotein and the P1 and P2 basic proteins. P1 appears to be identical to MBP, but P2 is distinctly different. PLP and P0 proteins are readily classified as integral membrane proteins, and the basic proteins are, as a class, the peripheral proteins of the myelin membrane.

Although CNS myelin from most mammalian sources has the protein composition noted above, certain rodents have myelin that contains a second basic protein (Martenson et al., 1971). It is a protein that is homologous with MBP but has 40 residues deleted from the MBP sequence (Dunkley and Carnegie, 1974). A similar situation appears to exist in the PNS myelin of the species: the smaller basic protein is not like the P2 protein usually found in the myelin of other animals, but is related to MBP (Greenfield et al., 1980; Milek et al., 1981).

The P2 protein has been detected in all PNS myelins; however, its amount is very variable. In PNS myelins of guinea pig, mouse, and rat, P2 protein appears to be present in only minute amounts (Greenfield et al., 1980; Brostoff et al., 1980; Milek et al., 1981). Immunological techniques have had to be used to demonstrate its presence in the sciatic nerve myelins of these animals.

There have been several studies on the supramolecular organization of the protein components in the myelin membrane. Data from studies on CNS myelin architecture together with the previous biochemical and morphological studies produced a model for myelin component organization. These data and our current concepts have been recently reviewed (Braun, 1977; Benjamins and Morell, 1978; Boggs and Moscarello, 1978; Rumsby, 1978). Basically, it appears that segments of PLP are at the IPL while other portions of the molecule traverse the lipid bilayer. MBP is located at the MDL, partially buried in the bilayer (Jones and Rumsby, 1977) and interacting with adjacent MBP molecules to form dimers (Golds and Braun, 1978).

Research on the architecture of PNS myelin has also been conducted with studies showing the location of PO protein at the IPL and suggesting that the basic proteins (P1 and P2) are at the MDL (Peterson and Gruener, 1978). However, another study has localized P1 protein to the IPL by immunocytochemical techniques (Mendel and Whitaker, 1977). Also, reports which suggest a nonuniform distribution of the P2 protein within the myelin sheaths of PNS tissue (Trapp et al., 1979; De Armond et al., 1980) contrast with other data which suggest a uniform localization for P2 protein (Eylar et al., 1980). Thus, it is not clear that the model of myelin architecture, developed essentially from CNS studies, also applies to the PNS membrane.

Regardless of the present uncertainties, the studies to date on CNS and PNS myelins have produced a model for myelin architecture. Myelin appears to be a relatively simple membrane, having only a few types of major proteins localized as might be expected from the prevailing paradigm for the types of proteins found in membranes. The major proteins are ascribed the functions of stabilizing the multilamellar structure through protein-protein and protein-lipid interactions. However, the current model neither provides a thorough comparative analysis of the architecture in CNS and PNS myelins nor contains details of the native conformations assumed by major proteins in the membrane. The type and significance of protein-protein interactions remain ill-defined, and the data that provide reasons for the multiplicity of basic proteins are nonexistent.

Prospects for obtaining a complete understanding of myelin architecture are rather good since significant progress has been made toward complete characterizations of myelin protein components. Data are slowly becoming available for the sequences of PLP (Jolles et al., 1979) and PO protein (Ishaque et al., 1980), while the primary structures for the MBP (Eylar et al., 1971; Carnegie, 1971) rat SBP (Dunkley and Carnegie, 1974), and P2 protein (Kitamura et al., 1980; Eylar et al., 1981) have already been determined. The availability of a more complete picture for the basic proteins derives from the facts that they are easier to work with and that the major CNS and PNS basic proteins, MBP and P2 protein, have been identified as molecules capable of inducing experimental demyelinating diseases in animals.

The data regarding the structure of MBP have been reviewed by Braun and Brostoff (1977). A schematic of MBP is shown in Fig. 1. The MBP molecule is comprised of 169 residues, has a high content of basic amino acids, and consequently has a high isoelectric point. In solution, MBP assumes a largely random conformation and it is thought to form a hairpin structure. The molecule contains an arginine residue which can be unmodified or mono- or di-methylated and a serine residue which is the primary location for phosphorylation of the protein at a level of one phosphate per five molecules. These modifications produce a certain amount of microheterogeneity in MBP. Whether MBP microheterogeneity has any functional significance remains to be determined. Because of the extensive work on MBP, it is one of the best characterized membrane proteins and the first membrane to have its complete sequence determined.

Several reports (Ishaque et al., 1980; Weise et al., 1980; Hsieh et al., 1981) have presented data regarding the primary structure of P2 protein, and the complete sequence of that protein is now known (Kitamura et al., 1980; Eylar et al., 1981). Fig. 2 contains a schematic representation of P2 protein structure. It is smaller than MBP, having only 131 amino acids, but is also acetylated at the amino terminal. There is no sequence homology with MBP. P2 protein is unique in that it contains two half cysteine residues per molecule which are capable of forming an intrachain disulfide bond. The conformation of P2 in solution is characterized by a large (60%-80%) component of beta-structure (Brostoff et al., 1975; Uyemura et al., 1977; Thomas et al., 1977; Deibler et al., 1978). An analysis of the primary sequence for conformation potential (Martenson, 1981) suggests that this strong preference for beta-structure arises as the result of long range interactions produced by folding the polypeptide backbone. A predicted secondary structure for P2 protein contains a series of beta strands separated by relatively short regions of coil or beta turns. A short segment of alpha helix is predicted to exist near

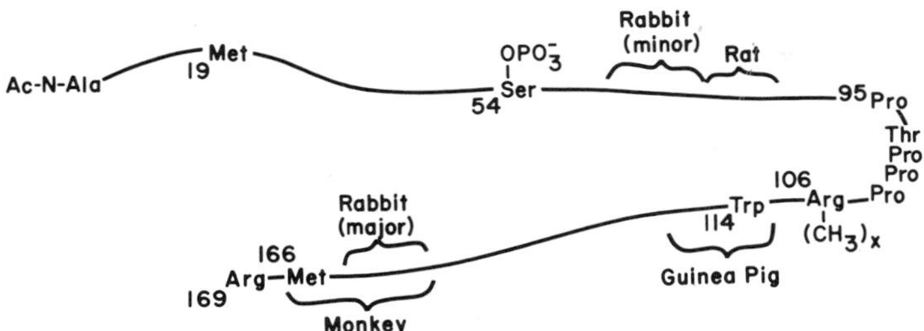

Fig. 1. Schematic of the structure for Myelin Basic Protein and relative location of encephalitogenic domains. Braun and Brostoff (1977) have reviewed the data on MBP structure and salient features of the molecule are shown in the schematic. Disease-inducing regions of the protein refer to bovine MBP in all cases except for Lewis rat where guinea pig protein was studied (Eylar, 1980).

the middle of the backbone. The schematic is not intended to indicate aspects of secondary structure.

Like MBP, CNBr digestion of the P2 protein removes short segments at both amino and carboxyl terminals of the and produces a large peptide from the central portion of the protein (Fig. 2, region B). In the case of P2, the large peptide does not appear to have a conformation with the high content of beta structure found in the parent molecule (Weise and Brostoff, 1981). It thus seems possible that the terminal segments of the sequence are important determinants of secondary structure for the P2 protein, and may well be involved in the long range interactions predicted in the analysis of conformational potential.

Both MPB and P2 protein are known to be able to produce experimental demyelinating diseases in animals. The inducible animal diseases of interest here are experimental allergic encephalomyelitis (EAE) and experimental allergic neuritis (EAN), affecting CNS and PNS tissues and functions, respectively (see Brostoff (1977) for a review of EAE and EAN). Clinical

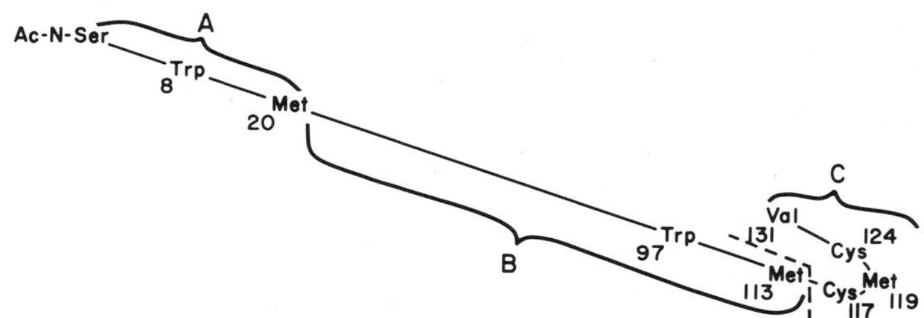

Fig. 2. The structure of P2 basic protein. The molecule is shown schematically with key amino acids. Segments A, B and C, the cyanogen bromide peptides of P2 protein, have been studied for their ability to induce experimental allergic diseases. See text for a discussion. Although the protein is known to have a high content of beta structure in solution, this is not represented in the figure. The only feature of molecular folding presented is at the carboxyl terminal where an intrachain disulfide can be found.

manifestations of EAE include paralysis, ataxia, and sphincter disturbances, while in EAN there is weakness of all extremities, facial weakness and respiratory involvement. EAE and EAN were initially induced by injection of the appropriate nervous tissue into animals. However, with extensive research, definition of the relevant antigens has been achieved. Animals contract EAE if sensitized with MBP in complete Freund's adjuvant (CFA) while EAN is observed in animals injected with P2 protein in CFA. Histologic lesions in experimental demyelinating diseases show the presence of perivascular infiltrates of mononuclear cells and demyelination with sparing of axons. Destruction of the myelin sheath seems to result from a stripping process mediated by macrophages.

The observed histological signs are consistent with a cell mediated immune response as the basis of EAE and EAN. In such a framework (Paterson, 1980), the first immunological event after introduction of the relevant antigen would be sensitization and activation of T lymphocytes. These would proliferate, migrate to tissue containing the target antigen, and there release lymphokines. These soluble factors would be directly responsible for initiating inflammation in the nervous tissue as macrophages accumulate. The destruction of myelin would result from the release of degradative enzymes from the activated macrophages.

Knowing that EAE and EAN are induced by the basic proteins and having determined the primary structures for those molecules, it is possible to further define the nature of the disease inducing structure. This can be done by obtaining peptide fragments of the parent proteins and testing them for disease-inducing potency. Unequivocal definition of the structure required for disease production can be achieved by utilizing synthetic peptides with sequences related to the disease-producing fragments.

II. ENCEPHALITOGENIC DOMAINS OF MBP

Over the past decade the above approach has been employed in delineating the encephalitogenic domains of MBP. Studies have utilized several species of test animals and species specific domains have been identified for induction of EAE in guinea pig, monkey, rabbit, and rat. The extensive data from that research have been reviewed by Eylar (1980) and are summarized below. Fig. 1 shows the relative location of encephalitogenic determinants for the animal species studied.

Guinea pig. Only one region in MBP appears capable of inducing EAE in guinea pigs. The relevant domain is a stretch of nine amino acids starting at residue 113 and including the only tryptophan found in the molecule. Extensive research with synthetic peptides has determined the amino acids in this region that are essential for disease production. The sensitivity of the induction process is seen in the fact that even a glycine for alanine substitution greatly affects disease inducing potency.

Monkey. The most potent region of MBP for inducing EAE in monkeys resides in the sequence between residues 153 and 166. Other minor determinants can be found in the amino terminal half of MBP. The tryptophan region of the molecule is inactive.

Rabbit. The first region to be identified as encephalitogenic in rabbits encompasses residues 64 to 73. However, through additional work with other peptides and chemically modified MBP, it was realized that the major determinant is located near MBP's carboxyl terminal, residues 153 to 166. This region is capable of producing both clinical and histological signs of EAE.

Lewis Rat. EAE induction in this species has proved to be a rather special case. Guinea pig MBP is much more active in producing EAE in Lewis rats than MBP from any other species tested. Extensive research has been required to establish that the encephalitogenic domain is in the region between residues 74 and 81 of MBP. In the course of that work, synthetic peptides were used to successfully explain the observed difference in encephalitogenicity of guinea pig and bovine MBPs.

The existence of various domains and differential susceptibility of the different animal species is presumably an indication that an immune response gene controls a particular species reaction to the different regions of the MBP molecule. The specific amino acid sequences of the encephalitogenic domains are not only recognized as foreign by the histocompatibility locus, but they probably also identify CNS myelin as a target for the sensitized T lymphocytes (Eylar, 1980). In this context the conformation of MBP and its disposition in the membrane are important.

III. NEURITOGENIC DOMAINS OF P2

The research effort to determine the neuritogenic domains in the P2 protein is relatively new. At the present time, the disease inducing properties of only the cyanogen bromide peptides of P2 have been studied. Disease induction studies can be summarized as follows.

Lewis Rat. Weise et al. (1980) have shown that EAN is readily induced with the large CNBr fragment that represents 70% of the entire P2 protein sequence and accounts for the central portion of the protein molecule. The sequences at either the amino or carboxyl terminals are unable to produce disease.

Pure bovine P2 protein has been used to induce disease in the Lewis rat (Kadlubowski and Hughes, 1979; Hoffman et al., 1980; Kadlubowski et al., 1980). The protein from rabbit, however, apparently requires additional factors to be active or to be maintained in an active conformation (Curtis et al., 1979). The sequence of the rabbit protein is reported to be different from that of bovine P2 at nine residues (Eylar et al., 1981). Seven of these differences are found in the neuritogenic portion of the molecule, with four of the seven being concentrated within a 16 residue segment containing trypotophan 97. This could be the reason for the differences in disease inducing properties for the rabbit and bovine P2 proteins.

Rabbit. EAN is not routinely induced in rabbits by injection of P2 protein in CFA. However, the CNBr fragments of the protein do induce disease as judged by histological signs, suggesting that protein conformation may be an important factor in the rabbit (Brostoff et al., 1977). The carboxyl terminal portion of P2, represented as a pair of small peptides linked by a disulfide bond, is the most consistent in producing EAN (Hsieh et al., 1981). Rabbits do appear to respond to other regions of the P2 molecule but this is less well studied.

Guinea pig. This species responds to challenge with PNS antigens but the resultant pathology is not confined to PNS tissue (Brostoff et al., 1973) as is the case in studies with both the rabbit and Lewis rat. Early studies detected a sensitivity to P2 protein (Carlo et al., 1975) and recent work has found that the amino terminal CNBr peptide from P2 protein induces an EAE-like disease in guinea pigs (Brostoff et al., 1980).

IV. DISCUSSION AND CONCLUSION

If the current line of research into defining the neuritogenic domains of P2 protein is compared to the past history with MBP, a similar progression might be expected. At some time in the future, it is likely that defined sequences along the P2 molecule will have been identified as domains that induce EAN. However, since the P2 molecule has considerable secondary structure and the potential for existing in states with either free sulfhydryls or an intrachain disulfide bond, it is also possible that there are determinants which result from the juxtaposition, in the folded molecule, of amino acid sequences from different regions of the primary structure. Judging from the studies in rabbits and rats, and from the apparent difference in conformation between P2 and its large CNBr peptide, this may not occur to any extent, but there are as of yet no studies that explore the potential of conformation dependent determinants.

In our efforts to understand EAE and EAN, there are three basic areas of research. Studies with MBP and P2 protein, known inducing agents for experimental demyelination, serve to define the segments of molecules important for disease induction. Examination of the immune response in sensitized animals leads us to the controlling factors and modes of immune reaction that produce demyelination. Finally, studies of myelin architecture are important for knowledge regarding the way(s) that nervous tissue is recognized as a target. With efforts in these three areas, the ultimate goal is to derive information that allows for an understanding of diseases like multiple sclerosis and Guillain-Barre syndrome with the possibility of preventing them or producing remission at the first signs of their onset.

V. REFERENCES

Benjamins, J. A. and Morell, P. (1978). Proteins of myelin and their metabolism. *Neurochem. Res.*, 3, 137-174.

Boggs, J. M. and Moscarello, M. A. (1978). Structural organization of the human myelin membrane. *Biochim. Biophys. Acta.*, 515, 1-21.

Braun, P. E. (1977). Molecular architecture of myelin. In "Myelin." (P. Morell, ed.) pp. 91-116. *Plenum, N.Y.*

Braun, P. E. and Brostoff, S. W. (1977). Proteins of myelin. In "Myelin." (P. Morell, ed.) pp. 201-232. *Plenum, N.Y.*

Brostoff, S. W. (1977). Immunological responses to myelin and myelin components. In "Myelin." (P. Morell, ed.) pp. 415-446. *Plenum, N.Y.*

Brostoff, S. W., Levit, S., and Powers, J. M. (1977). Induction of experimental allergic neuritis with a peptide from myelin P2 basic protein. *Nature*, 268, 752-753.

Brostoff, S. W., Powers, J. M., and Weise, M. J. (1980). Allergic encephalomyelitis induced in guinea pigs by a peptide from the NH2 terminus of bovine P2 protein. *Nature*, 285, 103-104.

Brostoff, S. W., Sacks, H., and Dipaola, C. (1975). The P2 protein of bovine root myelin: partial chemical characterization. *J. Neurochem.*, 24, 289-294.

Brostoff, S. W., Weise, M. J., and Powers, J. M. (1980). Allergic encephalomyelitis in guinea pigs induced by a peptide from the NH2-therminus of bovine P2 protein. *Trans. Am. Soc. Neurochem.*, 11(1), 133.

Brostoff, S. W., Wisniewski, H. M., Greenfield, S., Morell, P., and Eylar, E. H. (1973). Immunopathologic response in guinea pigs sensitized with peripheral nervous system myelin. *Brain Res.*, 58, 500-505.

Carlo, D. J., Karkhanis, Y. D., Bailey, P. J., Wisniewski, H. M., and Brostoff, S. W. (1975). Experimental allergic neuritis. Evidence for the involvement of the P0 and P2 proteins. *Brain Res.*, 88, 580-584.

Carnegie, P. R. (1971). Amino acid sequence of the encephalitogenic basic protein from human myelin. *Biochem. J.*, 123, 57-67.

Curtis, B. M., Forno, L. S., and Smith, M. E. (1979). Reactivation of neuritogenic activity of P2 protein from rabbit PNS myelin. *Brain Res.*, 175, 387-391.

De Armond, S. J., Deibler, G. E., Bacon, M., Kies, M. W., and Eng, L. F. (1980). A neurochemical and immunocytochemical study of P2 protein in human and bovine nervous systems. *J. Hist. Cyto-chem.* 28, 1275-1285.

Deibler, G. E., Driscoll, B. F., and Kies, M. W. (1978). Immunochemical and biochemical studies demonstrating the identity of a bovine spinal cord protein (SCP) and a basic protein of bovine peripheral nerve myelin (BF). *J. Neurochem.*, 30, 401-412.

Dunkley, P. R. and Carnegie, P. R. (1974). Amino acid sequence of smaller basic protein from rat brain myelin. *Biochem. J.*, 141, 243-255.

Eylar, E. H. (1980). The induction and suppression of experimental allergic encephalomyelitis. In "The suppression of experimental allergic encephalomyelitis and multiple sclerosis." (A. N. Davison and M. L. Cuzner, eds.) pp. 11-30. *Academic Press, N.Y.*

Eylar, E. H., Brostoff, S. W., Hashim, G., Caccam, J., and Burnett, P. J. (1971). Basic A1 protein of the myelin membrane. The complete amino acid sequence. *Biol. Chem.*, 246, 5770-5784.

Eylar, E. H., Ishaque, A., and Hofmann, T. (1981). The complete sequence of the rabbit P2 protein of PNS myelin. *Trans. Am. Soc. Neurochem.*, 12(1), 100.

Eylar, E. H., Szymanska, I., Ishaque, A., Ramwani, J., and Dubiski, S. (1980). Localization of the P2 protein in peripheral nerve. *J. Immunol.*, 124(3), 1086-1092.

Golds, E. E. and Braun P. E. (1978). Protein associations and basic protein conformation in the myelin membrane. *J. Biol. Chem.*, 253, 8162-8170.

Greenfield, S., Brostoff, S. W., and Hogan, E. L. (1980). Characterization of the basic proteins from rodent peripheral nervous system myelin. *J. Neurochem.*, 34, 453-455.

Hoffman, P. M., Powers, J. M., Weise, M. J., and Brostoff, S. W. (1980). Experimental allergic neuritis in the rat strain differences in the response to bovine P2 protein. *Brain Res.*, 195, 355-362.

Hsieh, D. L., Weise, M. J., Levit, S., Powers, J. M., and Brostoff, S. W. (1981). Structure of bovine P2 basic protein: Sequence of a carboxylterminal segment that is a neuritogen in rabbits. *J. Neurochem.*, 36, 913-916.

Ishaque, A., Hoffman, T., Rhee, S., and Wylar, E. H. (1980). The NH2-terminal region of the P2 protein from rabbit sciatic nerve myelin. *J. Biol. Chem.*, 255, 1058-1063.

Ishaque, A., Roomi, M. W., Szymanska, I., Kowalski, S., and Eyler, E. H. (1980). The PO glycoprotein of peripheral nerve myelin. *Canad. J. Biochem.*, 58, 913-921.

Jolles, J., Schientgen, F., Jolles, P., Vacher, M., Nocot, C., and Alfsen, A. (1979). Structural studies of the apoprotein of the Folch-PI bovine bovine brain myelin proteolipid: Characterization of the CNBr-fragments and of a long C-terminal sequence. *Biochem. Biphys. Res. Comm.*, 87, 619-626.

Jones, A. J. S. and Rumsby, M. G. (1977). Localization of sites for ionic interaction with lipid in the C-terminal third of the bovine myelin basic protein. *Biochem. J.*, 167, 583-591.

Kadlubowski, M. and Hughes, R. C. A. (1979). Identification of the neuritogen for experimental allergic neuritis. *Nature*, 277, 140-141.

Kadlubowski, M., Hughes, R. C. A., and Gregson, N. A. (1980). Experimental allergic neuritis in the Lewis rat: Characterization of the activity of peripheral myelin and its major basic protein. *Brain Res.*, 184, 439-454.

Kitamura, K., Suzuki, M., Suzuki, A., and Uyemura, K. (1980). The complete amino acid sequence of the P2 protein in bovine peripheral nerve myelin. *FEBS. Lett.*, 115, 27-30.

Martenson, R. E. (1981). Predicted structure of the myelin P2 protein. *Trans. Am. Soc. Neurochem.*, 12(1), 121.

Martenson, R. E., Deibler, G. E., and Kies, M. W. (1971). The occurrence of two myelin basic proteins in the central nervous system of rodents in the suborders *Myomorpha* and *Sciuromorpha*. *J. Neurochem.*, 18, 2427-2433.

Mendel, J. R. and Whitaker, J. N. (1977). Immunocytochemical localization studies of myelin basic protein. *J. Cell Biol.*, 76, 502-511.

Milek, D. J., Sarvas, H. O., Greenfield, S., Weise, M. J., and Brostoff, S. W. (1981). An immunological characterization of the basic proteins of rodent sciatic nerve myelin. *Brain Res.* (in press).

Norton, W. T. (1977). Isolation and characterization of myelin. In "Myelin." (P. Morell, ed.) pp. 161-200. *Plenum, N.Y.*

Paterson, P. Y. (1980). The immunopathology of experimental allergic encephalomyelitis. In "The suppression of experimental allergic encephalomyelitis and multiple sclerosis." (A. N. Davison and M. L. Cuzner, eds.) pp. 11-30. *Academic Press, N.Y.*

Peterson, R. B., and Gruener, R. W. (1978). Morphological localization of PNS myelin proteins. *Brain Res.*, 152, 17-29.

Raine, C. S. (1977). Morphological aspects of myelin and myelination. In "Myelin." (P. Morell, ed.) pp 1-50. *Plenum, N.Y.*

Rumsby, M. C. (1978). Organization and structure in central-nerve myelin. *Biochem. Soc. Trans.*, 6, 448-462.

Thomas, W. H., Weser, U., and Hempel, K. (1977). Conformational changes induced by ionic strength and pH in two bovine myeline basic proteins. *Hoppe-Seyler's Z. Physiol. Chem.*, 358, 1345-1352.

Trapp, B. D., McIntyre, L. J., Quarles, R. H., Sternberger, N. H., and Webster, H. deF. (1979). Immunochemical localization of rat peripheral nervous system myelin proteins: P2 protein is not a component of all peripheral nervous system myelin sheaths. *Proc. Natl. Acad. Sci. USA*, 79, 3552-3556.

Uyemura, K., Kato-Yamanaka, T., and Kitamura, K. (1977). Distribution and opical activity of the basic protein in bovine peripheral nerve myelin. *J. Neurochem.*, 29, 61-69.

Weise, M. J. and Brostoff, S. W. (1981). Determinants of secondary structure for nerve root P2 protein. *Trans. Am. Soc. Neurochem.*, 12(1), 101.

Weise, M. J., Hsieh, D. L., Hoffman, P. M., Powers, J. M., and Brostoff, S. W. (1980). Bovine peripheral nervous system P2 protein: Chemical and immunochemical characteristics of the cyanogen bromide peptides. *J. Neurochem.*, 35, 393-399.

Weise, M. J., Hsieh, D. L., Levit, S., and Brostoff, S. W. (1980). Bovine P2 protein: Sequence at the NH2-terminal of the protein. *J. Neurochem.*, 35, 388-392.

ORGANIZATION OF LYMPHOCYTE MEMBRANE PROTEINS

Michael J. Crumpton, Adelina A. Davies, Michael J. Owen, and Noel M. Wigglesworth

Imperial Cancer Research Fund
Lincoln's Inn Fields
London WC2A 3PX
United Kingdom

I. INTRODUCTION

I want to make three points by way of an introduction. Firstly, the lymphocyte surface plays crucial roles in regulating lymphocyte behavior. Most importantly, the lymphocyte surface mediates antigen recognition. As well, the surface membrane regulates the immediate biochemical consequences of antigen recognition that lead ultimately to lymphocyte growth, division, and differentiation into immunocompetent cells (e.g., antibody-secreting, T-killer and memory cells). The lymphocyte surface also mediates interaction with other cells, such as between killer and target cells, and of helper/suppressor cells with B lymphocytes and/or activated macrophages. In the latter respects, the gene products of the major histocompatibility region are especially important. Thus, as shown originally by Zinkernagel and Doherty (1979), the classical major histocompatibility antigens (H-2D, K, and L antigens in mice; HLA-A, B, and C antigens in humans) participate in the recognition and interaction of T-killer cells with virally-infected cells, whereas Ia antigens (I-region associated antigens in mice; HLA-DR antigens in humans) play a similar role in the interactions between immunocompetent cells.

My second point concerns the role of intracellular cytoskeletal elements in cell surface events. There is increasing evidence that a variety of lymphocyte surface events are regulated intracellularly by the cytoskeleton (Edelman, 1976). Such events include "patching and capping" of surface components in response to the binding of multivalent ligands, the internalization of receptor-ligand complexes, plasma membrane recyclization and, conceivably, lymphocyte-cell interaction. The putative relationship between the lymphocyte surface and the internal cytoskeleton is consistent with Keith Porter's studies on the structure of the cytoplasm of nucleated cells using the one million electron volt microscope (Wolosewick and Porter, 1979). Thus, Porter conceives of the inner surface of the plasma membrane of nucleated cells as being continuous with a fibrillar network of microtrabeculae that links the cytoplasmic face of the membrane to microtubules, microfilaments, endoplasmic reticulum, mitochondria, and the nucleus. It appears conceivable that this microfibrillar network is the nucleated cell's counterpart of the spectrin-actin-ankyrin complex that underlies the surface membrane in erythrocytes (Hainfield and Steck, 1977; Scheetz and Sawyer, 1978; Lux, 1979) and that it provides the means by which the cytoskeleton regulates the lymphocyte surface structure.

My third point is that if we are to understand at the molecular level how the lymphocyte surface regulates lymphocyte behavior, then we must satisfy the following requirements. Firstly, it is important to identify, isolate, and characterize the specific surface receptors (the antigen receptor of T- and B-lymphocytes, the major histocompatibility antigens, etc.). It is, however, also important to designate the nature of the signal that is generated at the lymphocyte surface by receptor-ligand (antigen) interaction and how this information is transmitted across the surface membrane into cell interior. Because the nature of the signal and the transmission mechanism are probably determined by the mode of association of the receptor with the lipid bilayer, there is a requirement to determine whether the receptor is a peripheral or integral membrane protein and whether it has a transmembrane orientation. Also, because the disposition and redistribution of surface components are probably regulated by the cytoskeleton via its interaction with the inner surface of the plasma membrane, there is a requirement to define the structure of the inner membrane surface and its mode of association with the cytoskeleton.

My laboratory is concerned with exploring various of these aspects. Today, I am going to review two of our recent studies. Firstly, I will describe a method to distinguish integral from peripheral membrane proteins and in particular to determine whether the antigen-receptor of B-lymphocytes (the subunit of IgM and/or IgD) dips into the lipid bilayer (also, see especially Kehry et al., 1980; and Rogers et al., 1980). Secondly, I will explore the possibility that the lymphocyte plasma membrane has attached to its inner surface a fibrillar network similar to that of erythrocytes.

II. DESIGNATION OF INTEGRAL MEMBRANE PROTEINS

In addition to what is presented below, see Owen et al. (1980).

Our approach to the designation of integral membrane proteins is, in principle, identical with those of previous workers in this field (Klip and Gitler, 1974; Bayley and Knowles, 1978a,b; Bercovici and Gitler, 1978), namely, to design a highly lipophilic reagent that reacts indiscriminately with polypeptides. Such a reagent should label only those polypeptides in contact with the hydrocarbon phase of the lipid bilayer. The reagent we designed is shown in Fig. 1. It consists of an aromatic azide residue which can be activated by light to form a highly reactive nitrene. The aromatic azide is attached to a tyramine residue which can be labeled with radioactive iodine to high specific activity and which is also acylated with a hexanoyl group in order to increase its hydrophobicity. An important question is, how lipophilic is the reagent? Partition experiments using octanol and water indicate that it has a partition coefficient of at least 10^4:1 and, thus, that it is extremely lipid soluble. A more important question is, however, whether it labels only integral membrane polypeptides? This question was explored using the human classical major transplantation antigens (HLA-A, B antigens) as a model. Fig. 2 shows that the structure of these membrane glycoproteins has been extensively characterized. Thus, they comprise a glycosylated polypeptide of about 43,000 molecular weight that, on the basis of its amino acid sequence, vectorial labelling of the inner membrane surface and enzymic digestion spans the lipid bilayer (Springer and Strominger, 1976; Walsh and Crumpton, 1977). This polypeptide is noncovalently associated with a 12,000 molecular weight nonglycosylated polypeptide, namely β_2-microglobulin, that is extrinsic to the lipid bilayer. Given this structure, the reagent should label the 43,000 molecular weight but not the 12,000 molecular weight polypeptide. Further, as indicated in Fig. 2, the 43,000 molecular weight polypeptide is cleaved by papain at a position external to the lipid bilayer, thus releasing a water-soluble fragment; for obvious reasons, this fragment should not be labeled by the reagent.

The first experiment was to add the lipophilic aromatic azide to a purified preparation of the plasma membrane of human BRI 8

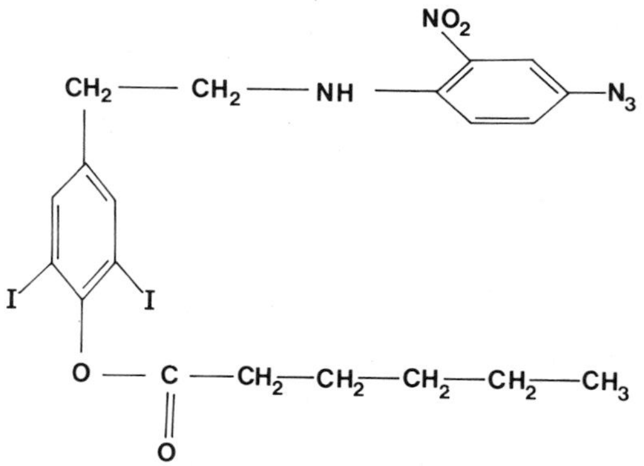

Fig. 1. Hexanoyldiiodo-N-(4-azido-2-nitrophenyl) tyramine.*

*Conversion to the reactive nitrene was achieved by photolysis at 480 nm.

lymphoblastoid cells and to generate the reactive nitrene by photolysis. The labeled membrane was next solubilized using the detergent Na deoxycholate and the HLA-A, B antigens precipitated by using a rabbit antiserum against human β_2-microglobulin (identical results were obtained using monoclonal antibodies against HLA-A, B antigens or β_2-microglobulin).

The control experiment comprised immunoprecipitation of Na deoxycholate-solubilised BRI 8 plasma membrane that had been labeled by lactoperoxidase-catalyzed iodination. The results are shown in Fig. 3. Clearly both the 43,000 and 12,000 molecular weight polypeptides were labelled in the control (Fig. 3B). In contrast, although the 43,000 molecular weight chain was labeled by the nitrene, no radioactivity was detected in the position of β_2-microglobulin (Fig. 3A). The effect of papain digestion on the distribution of radioactivity in nitrene-labeled HLA-A, B antigens is shown in Fig. 4. In this experiment a purified glycoprotein fraction comprising a mixture of HLA-A, B, and HLA-DR antigens (Fig. 4, track C) was labeled with the nitrene in the presence of detergent (Fig. 4, track A). Papain digestion of the detergent-solubilized, nitrene-labeled glycoprotein fraction resulted in the almost complete cleavage of the 43,000 weight polypeptide of the HLA-A, B antigens to give a chain of about 36,000 molecular weight (Fig. 4, track D). As shown in Fig. 4, track B, this chain is not associated with any radioactivity; further, after papain digestion, the radioactivity is located in a position

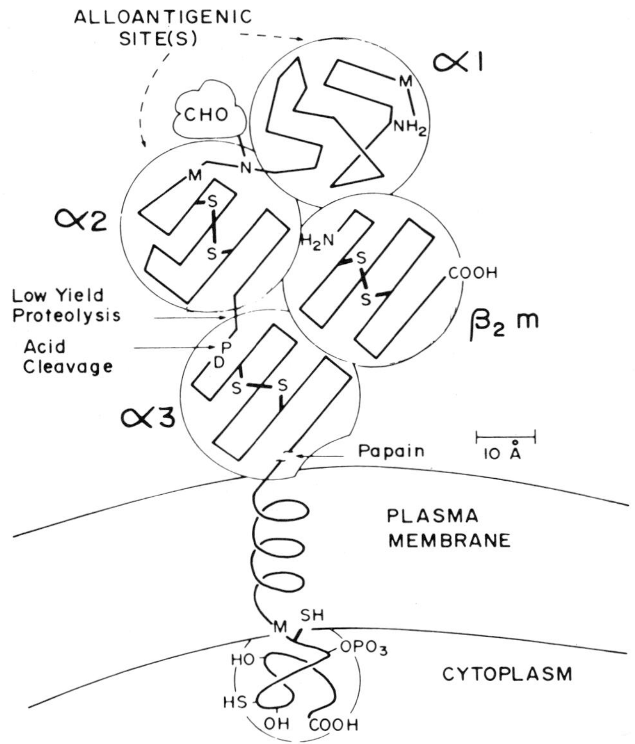

Fig. 2. Schematic representation of a molecule of an HLA-A or B antigen in the plasma membrane. The molecule is viewed as comprising four domains at the external surface of the lipid bilayer. β_2-Microglobulin comprises one domain; α_1, α_2, and α_3 domains are formed by the heavy chain. A hydrophobic stretch of the heavy chain traverses the lipid bilayer and a hydrophilic domain is exposed on the cytoplasmic face of the membrane. Digestion of the membrane associated or detergent-solubilized HLA-A, B antigens with papain produces a water-soluble fragment of apparent molecular weight 34,000 (See Fig. 3). Reproduced from Strominger et al., in: *Current Topics in Developmental Biology, Developmental Immunology*. Vol. 14, Acad. Press, New York. (in press).

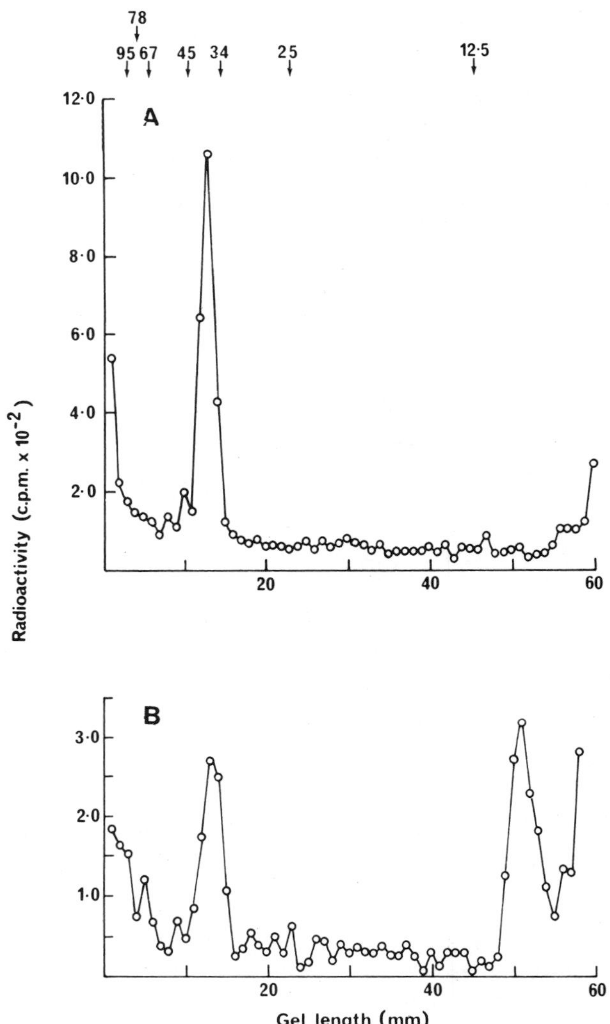

Fig. 3. Immunoprecipitation of BRI 8 plasma membrane with rabbit anti-(β_2-microglobulin) serum. Plasma membrane was radioiodinated by either the nitrene (A) or lactoperoxidase-catalyzed iodination (B) and the immunoprecipitates were analyzed by electrophoresis on a 15% polyacrylamide gel in N-dodecylsulfate under reducing conditions. The gel was stained, dried and divided into 1 mm sections prior to counting. Reproduced from Owen et al. (1980).

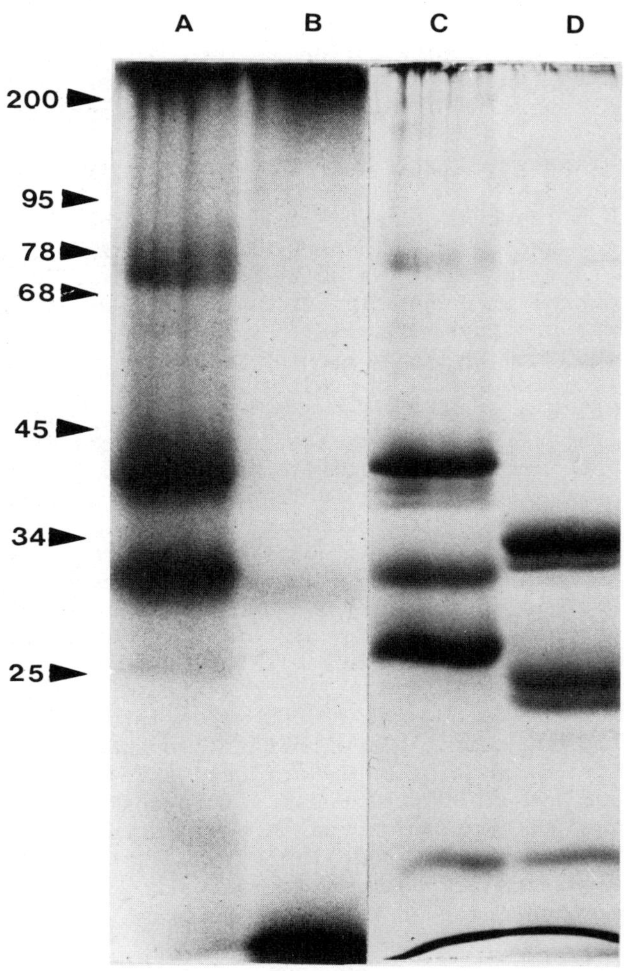

Fig. 4. Papain cleavage of a Na deoxycholate-solubilized, purified glycoprotein fraction containing HLA-A, B, and HLA-DR antigens labeled by the nitrene. Tracks A and B represent autoradiographs of the uncleaved and papain-cleaved glycoprotein fractions. The Coomassie blue stained patterns are depicted in tracks C and D. Reproduced from Owen et al. (1980).

identical with or close to the Na dodecylsulfate front, i.e., equivalent to a low molecular weight polypeptide. These results are consistent with the predicted specificity of the reagent and argue strongly in support of the conclusion that labeling of the HLA-A, B antigens was restricted to the intralipid portion of the 43,000 molecular weight polypeptide. Given this result it is reasonable to employ the reagent to explore the modes of lipid association of several other lymphocyte surface antigens.

HLA-DR antigens (Fig. 5) comprise two glycosylated polypeptide chains of about 33,000 and 28,000 molecular weight both of which span the lipid bilayer as judged by vectorial labeling of the inner membrane surface and by proteinase K digestion of endoplasmic reticulum membrane vesicles (Walsh and Crumpton, 1977; Owen et al., 1981). The control immunoprecipitation of BRI8 plasma membrane labeled by lactoperoxidase-catalyzed iodination gave the expected pattern of 33,000 and 28,000 molecular weight polypeptides (Fig. 6B). Immunoprecipitation of the nitrene-labeled plasma membrane (Fig. 6A) revealed a strongly labeled 33,000 molecular weight band, whereas the radioactivity associated with the 28,000 molecular weight chain was barely discernible. This uneven distribution of radioactivity was also characteristic of the nitrene-labeled purified glycoprotein fraction (Fig. 4, track A). As a similar disproportionate distribution of activity was obtained with rat Ia antigens, it is probably a general property of these antigens. On papain digestion, both the 33,000 and 28,000 molecular weight polypeptides were degraded, fragments of about 23,000 molecular weight being formed (Fig. 4, track D) that possessed no discernible radioactivity (Fig. 4, track B). Although the uneven distribution of label between the 33,000 and 28,000 molecular weight polypeptides is superficially not compatible with the accepted model of HLA-DR antigen structure is most probably reflects shielding of the intralipid portion of the 28,000 molecular weight chain by the like portion of the 33,000 molecular weight polypeptide. The possibility that the nitrene is able to probe intralipid protein-protein association is particularly attractive since no method currently exists for investigating this aspect of membrane structure.

One of the major reasons for developing the hydrophilic nitrene reagent was to use it to probe the mode of association of lymphocyte surface immunoglobulins with the membrane. Labeling of the membrane-associated immunoglobulin of BRI 8 cells was assessed after reacting the purified plasma membrane fraction with the nitrene. As shown in Fig. 7A, one peak only was detected in the position of the μ heavy chain of human immunoglobulin M, but no radioactivity was discernible in the light chain position. In contrast, both chains were labeled by lactoperoxidase catalyzed iodination (Fig. 7B). Similar labeling patterns were obtained

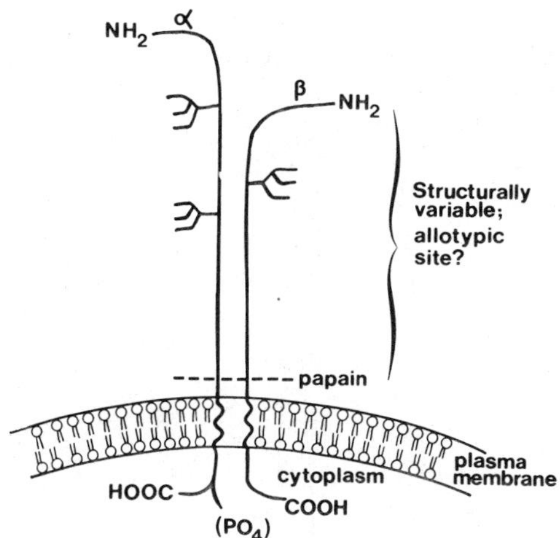

Fig. 5. Schematic representation of an HLA-DR molecule in the plasma membrane. The molecule is viewed as comprising two chains (α and β) which are tightly associated non-covalently and which span the lipid bilayer. Reproduced from Owen and Crumpton (1980).

with the plasma membrane fractions from Daudi cells and from mouse spleen cells except that in the latter case the nitrene labeled both the μ and δ heavy chains. These results, together with the apparent absence of labeling of human secreted immunoglobulin M in the presence of Triton X100, argue strongly that membrane immunoglobulins are associated directly with the lipid bilayer via their μ and δ chains. Further, they suggest that the heavy chains of membrane immunoglobulins differ structurally from those of secreted immunoglobulins in possessing an alternative or extra-hydrophobic C-terminal segment. Since these data were acquired, Dr. Lee Hood and his colleagues (Kehry et al., 1980; Rogers et al., 1980) have obtained direct evidence that the μ chains of membrane and secreted immunoglobulins M differ in their C-terminal segments and that in particular membrane immunoglobulin μ chains possess a hydrophobic region that spans the lipid bilayer. It is, thus, concluded that membrane immunoglobulin is an integral protein and that the previous suggestions which incorporated attachment to the lymphocyte surface via an Fc receptor are incorrect.

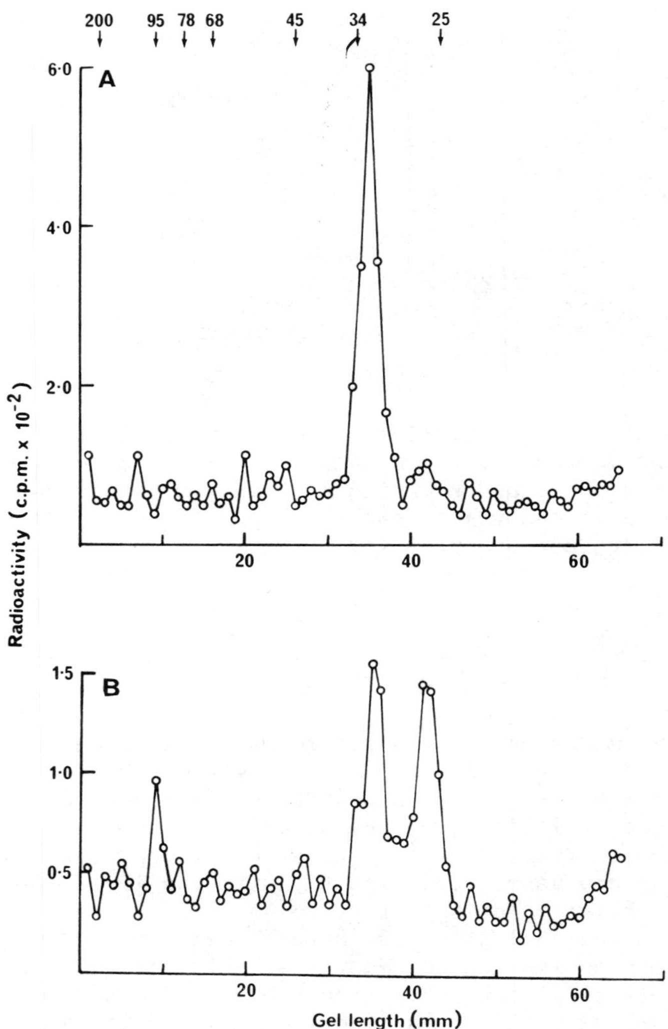

Fig. 6. Immunoprecipitation of BRI 8 plasma membrane with rabbit anti-(HLA-DR) serum. Plasma membrane was radioiodinated by either the nitrene (A) or lactoperoxidase-catalyzed iodination (B). Reproduced from Owen et al. (1980).

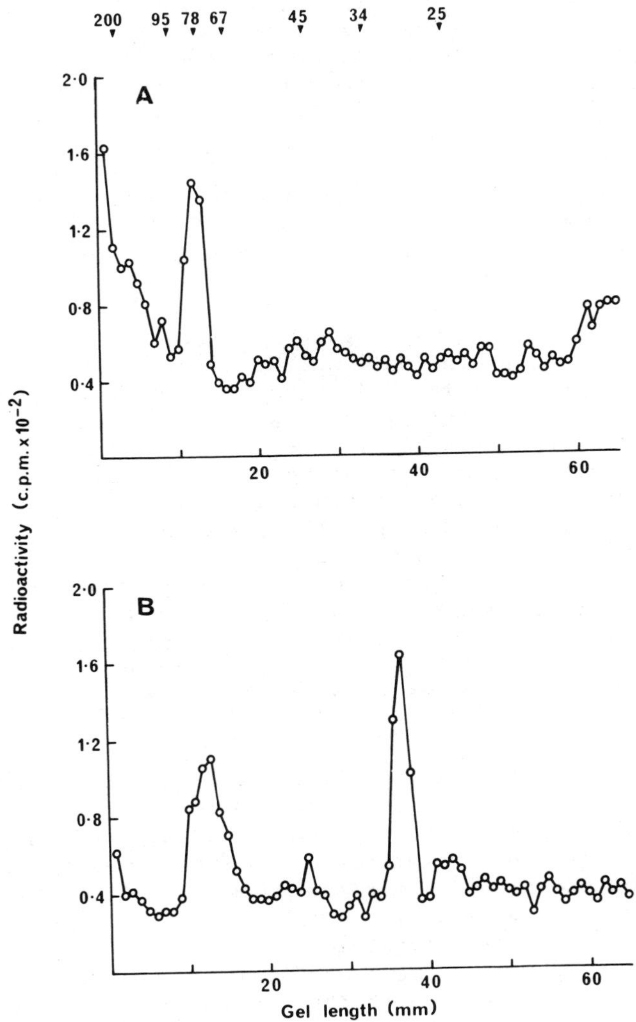

Fig. 7. Immunoprecipitation of BRI 8 plasma membrane with rabbit anti-(human Ig) serum. Plasma membrane was radioiodinated by either the nitrene (A) or lactoperoxidase-catalyzed iodination (B).

IV. LYMPHOCYTE PLASMA MEMBRANE SKELETON

The presence in lymphocyte plasma membrane of a fibrillar network resembling the spectrin-actin-ankyrin complex of erythrocytes was explored using the procedure illustrated in Fig. 8. This procedure is based upon the classical studies of Steck (Yu et al., 1973) who used a non-ionic detergent, namely Triton X100, to selectively solubilize (extract) the membrane lipids and proteins from erythrocyte ghosts. All of our studies have been carried out using cultured human B lymphobastoid cells; similar results were obtained using different cell lines (Daudi, BRI 8, Maja). Recently, Mescher et al. (1981) described the results of a similar study using the plasma membranes of murine tumor and lymphoid cells. The results of the two studies are in general agreement, although there are some notable differences.

After solubilizing with NP40 under physiological conditions i.e., pH, ionic strength, Ca^{2+} and Mg^{2+} concentration, about 12% of the protein of purified Maja plasma membrane was sedimented employing similar conditions to those used to separate the plasma membrane (microsome) fraction. Electron microscopy of the 20K pellet (Fig. 9) revealed some amorphous material plus, surprisingly, a fair proportion of membrane vesicles with the characteristic bilayer structure. Morphologically the vesicles were similar in size and appearance to those of the original plasma membrane preparation apart from a higher proportion of vesicles in contact with each other. Higher magnification revealed some myelin-like figures and an occasional trilaminar structure reminiscent of a hepatocyte tight junction. The significance of the latter structure is thought questionable, especially since the cells had been grown in suspension culture.

The presence of lipid in the 20K pellet was confirmed by analysis (Tables 1, 2). Lipid analyses further showed that the pellet resembled the original plasma membrane fraction in respect of the amount of lipid relative to protein, the ratio of phospholipid to cholesterol and the relative amounts of the different phospholipids. It was concluded that purified preparations of the plasma membrane of lymphoblastoid cells contain a significant amount of a population of membrane vesicles that in spite of having a similar lipid composition to the total membrane preparation is, nevertheless, insoluble in NP40. This result was totally unexpected and superficially, at least, is difficult to account for. One possible explanation is based upon the differential solubility in non-ionic detergents of the cell surface membrane compared with the nuclear membrane. Thus, when whole lymphocytes are "solubilized" in NP40, the cell surface membrane dissolves, but the nuclear membrane(s) and, in consequence, the nuclei remain intact. As a result it seems possible that the 20K pellet represents nuclear

Lymphocyte plasma membrane

0.5 mg protein/ml 0.5% NP40*

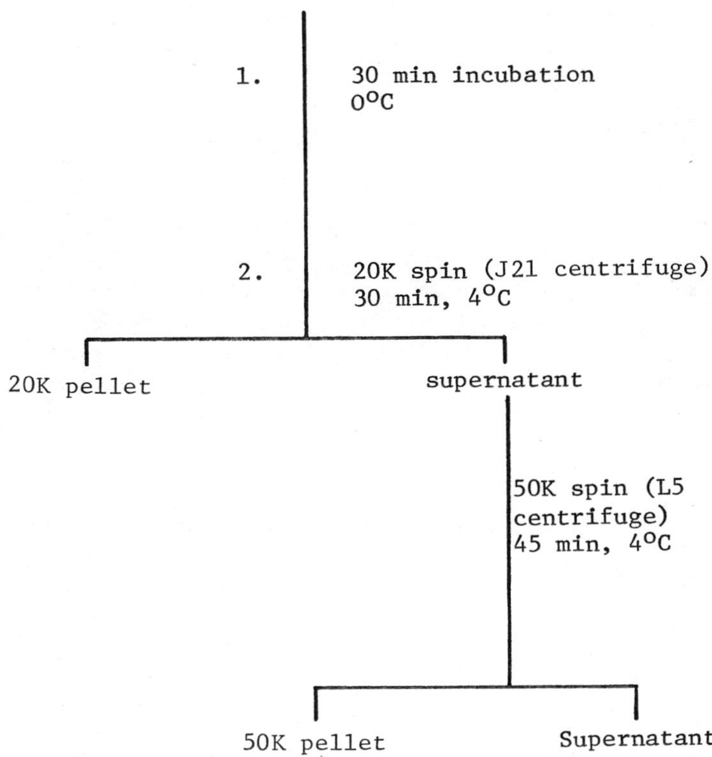

---

*NP40 dissolved in 10 mM phosphate buffer, pH 7.2, 0.17M NaCl, 3 mM KCl, 1 mM $CaCl_2$, 1 mM $MgCl_2$, 1 mM phenylmethanesulfonyl fluoride, 10 mM iodoacetamide.

Fig. 8. Procedure for "solubilizing" lymphocyte plasma membrane with non-ionic detergents.

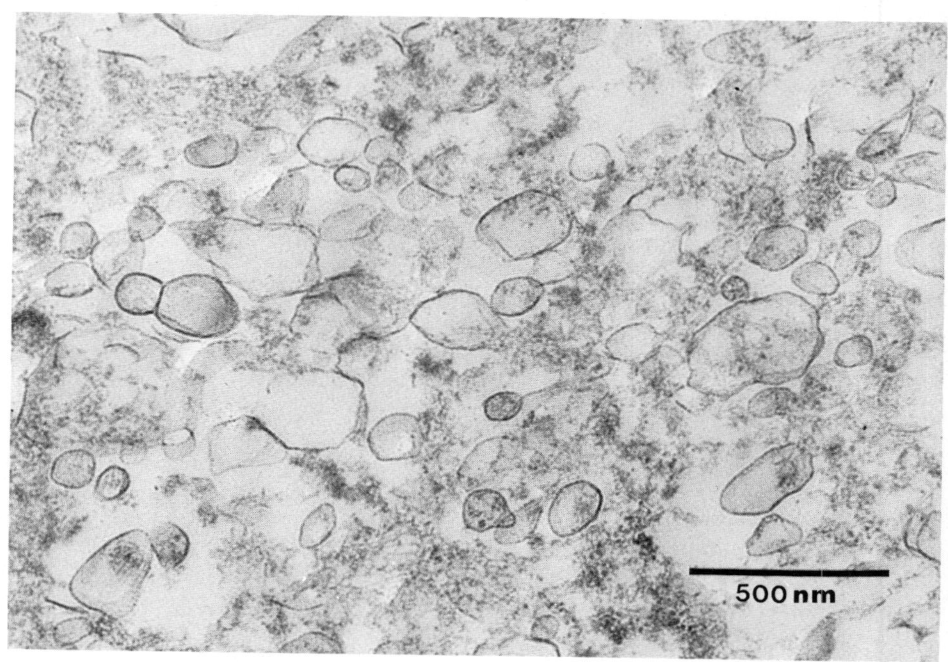

Fig. 9. Electron micrograph of the 20K pellet obtained after extracting purified BRI 8 plasma membrane with NP40. The micrograph was kindly prepared by Ms. Carol Upton.

membrane(s) that contaminated the original plasma membrane fraction. Two lines of evidence argue against this possibility. Firstly, although there is no authentic marker of nuclear membrane(s), glucose-6-phosphatase is generally accepted as a reasonable marker (also, of course, for endoplasmic reticulum). Table 3 shows that the purified Maja plasma membrane had a very low glucose-6-phosphatase activity and that this activity was not increased significantly in the 20K pellet. Incidentally, the 20K pellet contained no detectable 5'-nucleotidase activity (a plasma membrane marker) which is contrary to Mescher et al.'s (1981) experience. Secondly and more importantly, anti-(20K pellet) sera prepared in mice and rabbits failed to stain the nuclei of mouse fibroblasts and human lymphoblastoid cells (Fig. 11 and *vide infra*).

Table 1. Comparison of the lipid composition of purified lymphocyte plasma membrane (cell line; Daudi) with that of the 20K pellet obtained by NP40 extraction.*

Sample	Cholesterol (μmol/mg of protein)	Phospholipid (μmol/mg of protein)	Cholesterol:Phospholipid (molar ratio)
Plasma membrane	0.39	0.74	0.51
20K pellet	0.50	1.0	0.50

*Lipid analyses were kindly performed by Dr. D. Allan.

Table 2. Comparison of the phospholipid composition of purified lymphocyte plasma membrane (cell line; Daudi) with that of the 20K pellet obtained by NP40 extraction.*

Sample		Composition (%)					
		Expt. no. 1			Expt. no. 2		
Plasma membrane	PC		65 (PC + SM)		57	69 (PC + SM)	
	SM				12		
	PS	9	32 (PS + PE)			21 (PS + PE)	
	PE	23		36 (PS + PE + PI)			31 (PS + PE + PI)
	PI	4			10		
20K pellet	PC		76 (PC + SM)		58	77 (PC + SM)	
	SM				19		
	PS					14 (PS + PE)	
	PE			24 (PS + PE + PI)			23 (PS + PE + PI)
	PI						

*Lipid analyses were kindly performed by Dr. D. Allan.

The above results suggest that the membrane vesicles of the 20K pellet probably represent plasma membrane. In this case, given the currently accepted mechanism for membrane solubilization by non-ionic detergents (Helenius and Simons, 1975), it is difficult to account for the failure to solubilize the lipids unless this failure reflects the protein composition of the vesicles. The polypeptide composition of the 20K pellet of Maja plasma membrane is visualized in Fig. 10. The pattern is strikingly simple when compared with that of the soluble fraction (i.e., the 50K supernatant; Fig. 8), comprising three major polypeptides of 129,000,

Table 3. Glucose 6-phosphatase and 5'-nucleotidase activities of a purified preparation of lymphocyte plasma membrane (cell line; Maja) and of the NP40-insoluble 20K pellet

	Specific enzyme activity (μmol P_i released/mg protein/min)	
Enzyme	Plasma membrane	20K pellet
5'-nucleotidase	39	not detectable
Glucose 6-phosphatase	0.03	0.04

67,000, and 44,000 molecular weight together with about 10 minor bands of molecular weight varying from 200,000 to 30,000. The 129,000 and 67,000 molecular weight bands were located primarily, if not exclusively, in the 20K pellet but the 44,000 molecular weight band was present in both the 20K pellet and the 50K supernatant. As judged by co-migration with authentic proteins, the 200,000 and 44,000 molecular weight polypeptides probably represent myosin and actin, respectively, whereas, by analogy with similar solubilization studies carried out on the microvilli of intestinal epithelial cells, the 67,000 molecular weight may represent fimbrin (Bretscher and Weber, 1980). Fig. 10 also shows that the polypeptide composition of the 20K pellet can be altered dramatically by depleting the solubilizing solvent of Ca^{2+} and Mg^{2+}. Under these conditions, the 67,000 molecular weight band was located exclusively in the 50K supernatant.

The separation of a 20K pellet with the above restricted polypeptide composition was not exclusive to NP40 as the solubilizing detergent. Closely similar results were obtained using Triton X100 and sodium deoxycholate in phosphate buffered saline containing K^+, Mg^{2+}, and Ca^{2+}. In contrast, empigen BB gave a negligible sized 20K pellet that comprised few discernible polypeptides.

Antisera against the 20K pellet have been raised in mice and rabbits. As judged by immunofluorescence, these antisera did not bind to the lymphocyte surface but recognized intracellular structures in mouse 3T3 fibroblasts and human lymphoblastoid cells that had been fixed (paraformaldehyde) and permeabilized (acetone). The pattern given by the mouse antiserum was particularly striking (Fig. 11) in that it comprised bright patches of fluorescence located preferentially along the filopodia. It appears possible that these patches may correspond to the areas of attachment of the filopodia to the substratum. With human lymphoblastoid cells there was a fine

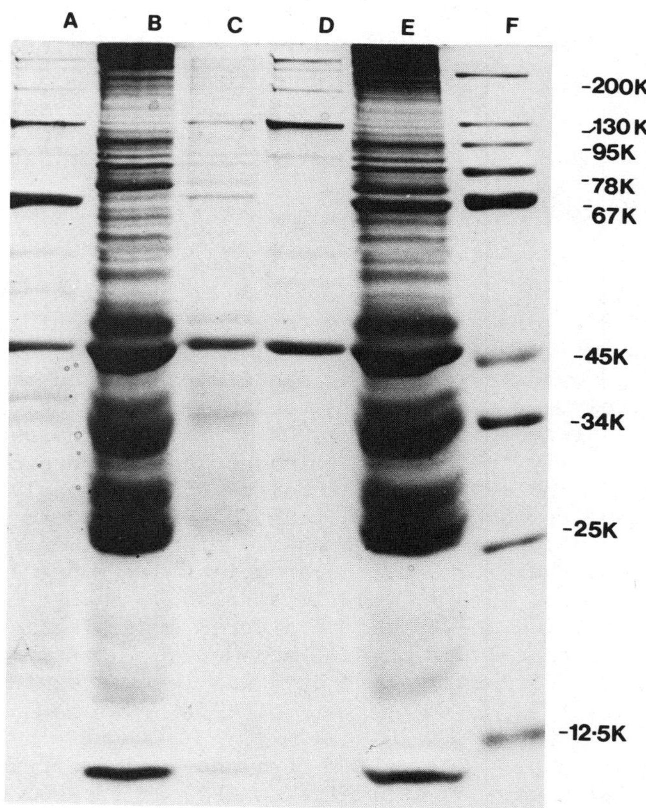

Fig. 10. Polyacrylamide gel electrophoretic analysis in Na dodecylsulfate. The gel contained 10% polyacrylamide and was stained with Coomassie blue. Tracks A and D represent 20K pellets obtained by extracting purified Maja plasma membrane with NP40 in the presence and absence respectively of Ca^{2+} and Mg^{2+}. Tracks B and E represent the corresponding 50K supernatants. Track F are standard proteins whose molecular weights are indicated on the right of the figure.

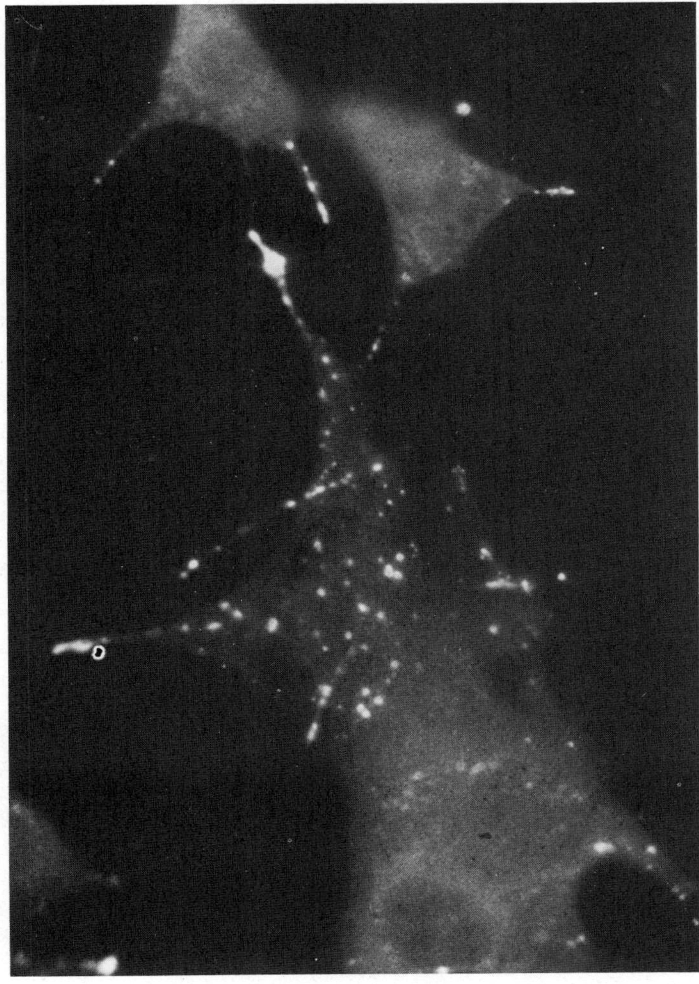

Fig. 11. Immunofluorescent staining by a mouse anti-(20K pellet from BRI 8 plasma membrane) serum of fixed, permeabilized mouse 3T3 fibroblasts. Antibody binding was revealed using a FITC-labeled goat anti-(mouse Ig) serum.

speckled pattern of fluorescence that appeared to be associated preferentially with the plasma membrane (by implication the membrane's inner surface since whole cells were not stained). It is important to know which polypeptides of the purified plasma membrane and 20K pellet are recognized by these antisera. Experiments are in progress to resolve this question as well as to prepare monoclonal antibodies against the major polypeptides of the 20K pellet.

To summarize the latter section of this presentation, the present studies indicate that about 10% of the protein of purified plasma membrane preparations from human lymphoblastoid cells is insoluble in various non-ionic detergents dissolved in Dulbecco's buffered saline. The insoluble (20K) pellet comprises, morphologically, amorphous material together with some membrane vesicles. These two components have, so far, not been separated. The overall lipid to protein ratio of the pellet was somewhat higher than that of the plasma membrane, but the lipid composition was remarkably similar to that of the plasma membrane. The pellet contained several minor polypeptide chains plus three major polypeptides of 129,000, 67,000, and 45,000 molecular weight. The two higher molecular weight chains were apparently located exclusively in the pellet fraction. If the detergent solubilization was carried out, in the absence of Ca^{2+} and Mg^{2+} then the 67,000 molecular weight polypeptide was contained in the soluble fraction. Antisera raised against the pellet reacted preferentially with intracellular components of mouse fibroblasts and human lymphoblastoid cells and apparently did not contain antibodies against the cell surface.

Obviously many questions remain to be explored including ones such as: Is all of the lipid associated with the membrane vesicles, do these vesicles contain all the polypeptides detected in the 20K pellet, and are these polypeptides directly responsible for the vesicles' insolubility in detergent? Also, do some of the polypeptides form a macromolecular network resembling the spectrin-actin-ankyrin complex of erythrocytes, what is the function(s) of the polypeptides of the detergent-insoluble fraction and, in particular, do they participate in or regulate lymphocyte surface-mediated phenomena such as patching, capping, and endocytosis? We await the answers to, at least, some of these questions with considerable interest.

IV. REFERENCES

Bayley H. and Knowles, J. R. (1978a). Photogenerated reagents for membrane labeling. 1. Phenylnitrene formed within the lipid bilayer. *Biochemistry*, 17, 2414-2419.

Bayley H. and Knowles, J. R. (1978b). Photogenerated reagents for membrane labeling. 2. Phenycarbene and adamantylidene formed within the lipid bilayer. *Biochemistry*, 17, 2420-2423.

Bercovici, T. and Gitler, C. (1978). 5-[^{125}I] Iodonaphthyl azide, a reagent to determine the penetration of proteins into the lipid bilayer of biological membranes. *Biochemistry*, 17, 1484-1489.

Bretscher, A. and Weber, K. (1980). Fimbrin, a new microfilament-associated protein present in microvilli and other cell surface structures. *J. Cell Biol.*, 86, 335-340.

Edelman, G. M. (1976). Surface modulation in cell recognition and cell growth. *Science*, 192, 218-226.

Hainfield, J. F. and Steck, T. L. (1977). The sub-membrane reticulum of the human erythrocyte: a scanning electron microscope study. *J. Supramol. Struct.*, 6, 301-311.

Helenius, A. and Simons, K. (1975). Solubilization of membranes by detergents. *Biochim. Biophys. Acta*, 415, 29-79.

Kehry, M., Ewald, S., Douglas, R., Sibley, C., Raschke, W., Fambrough, D., and Hood, L. (1980). The immunoglobulin μ chains of membrane-bound and secreted IgM molecules differ in the C-terminal segments. *Cell*, 21, 393-406.

Klip, A. and Gitler, C. (1974). Photoactivate covalent labeling of membrane components from within the lipid core. *Biochem. Biophys. Res. Commun.*, 60, 1155-1162.

Lux, S. E. (1979). Dissecting the red cell membrane skeleton. *Nature*, 281, 426-429.

Mescher, M. F., Jose, M. J. L., and Balk, S. P. (1981). Actin-containing matrix associated with the plasma membrane of murine tumour and lymphoid cells. *Nature*, 289, 139-144.

Owen, M. J. and Crumpton, M. J. (1980). Biochemistry of major human histocompatibility antigens. *Immunol. Today*, (Dec. 1980), 117-122.

Owen, M. J., Knott, J. C. A., and Crumpton, M. J. (1980). Labeling of lymphocyte surface antigens by the lipophilic, photoactivatable reagent hexanoyldiiodo-N-(4-azido-2-nitrophenyl) tyramine. *Biochemistry*, 19, 3092-3099.

Owen, M. J., Kissonerghis, A-M., Lodish, H. F., and M. J. Crumpton. (1981). Biosynthesis and maturation of HLA-DR antigens *in vivo*. *J. Biol. Chem.* (in press).

Rogers, J., Early, P., Carter, C., Calame, K., Bond, M., Hood, L., and Wall, R. (1980). Two mRNAs with different 3' ends encode membrane-bound and secreted forms of immunoglobulin μ chain. *Cell*, 20, 303-312.

Scheetz, M. P. and Sawyer, D. (1978). Triton shells of intact erythrocytes. *J. Supramol. Struct.*, 8, 399-412.

Springer, T. A. and Strominger, J. L. (1976). Detergent-soluble HLA antigens contain a hydrophilic region at the COOH-terminus and a penultimate hydrophobic region. *Proc. Nat. Acad. Sci.*, 73, 2481-2485.

Walsh, F. S. and Crumpton, M. J. (1977). Orientation of cell-surface antigens in the lipid bilayer of lymphocyte plasma membrane. *Nature*, 269, 307-311.

Wolosewick, J. J., and Porter, K. R. (1979). Microtrabecular lattice of the cytoplasmic ground substance: artifact or reality. *J. Cell Biol.*, 82, 114-139.

Yu, J., Fichman, D. A., and Steck, T. L. (1973). Selective solubilization of proteins and phospholipids from red blood cell membranes by nonionic detergents. *J. Supramolec. Struct.*, 1, 233-248.

Zinkernagel, R. M. and Doherty, P. C. (1979). MHC restricted cytotoxic T cells. Studies on the biological role of polymorphic major transplantation antigens determining T-cell restriction specificity function and responsiveness. *Adv. Immunol.*, 27, 51-177.

ANTIBODY GENES: ORGANIZATION, REARRANGEMENT AND DIVERSIFICATION

Leroy Hood

Division of Biology
California Institute of Technology
Pasadena, California 91125

The antibody molecules are encoded by three gene families--two encode light (L) chains, λ and κ, and the third encodes heavy (H) chains. The light chains are encoded by three distinct gene segments, variable (V_L), joining (J_L), and constant (C_L), whereas the heavy chains are encoded by four gene segments--V_H, diversity (D), J_H, and C_H (Fig. 1). In the mouse there are eight distinct C_H genes which encoded the various classes and subclasses of antibody molecules, e.g., Cμ-IgM, Cλ-IgG, and C_α-IgA.

Two types of DNA rearrangements occur during the differentiation of antibody-producing (B) cells-variable region formation which joins the V_L and J_L or V_H, D, and J_H gene segments and class or C_H switching which permits the initially expressed C_μ gene (IgM molecule) to be replaced by a second C_H gene (Fig. 1). Variable region formation appears to be mediated by recognition sequences which lie to the 3' sides of the V and D gene segments and to the 5' sides of the D and J gene segments (Early et al., 1980a). Presumably specific "joining" proteins recognize these sequences, form heterodimers, thus juxtaposing the adjacent gene segments which are then joined directly by a site-specific recombinational mechanism. Class switching probably occurs between homologous switch sequences lying to the 5' side of each C_H gene (Davis et al., 1980).

RNA splicing mechanisms lead to the simultaneous expression of four different types of heavy chains--secreted μ (μ_s), membrane μ (μ_m), secreted δ (δ_s), and membrane δ (δ_m). Because the C_μ and C_α genes are only 2.5 kilobases apart, a single RNA nuclear transcript can include the V_H, C_μ, and C_δ genes and be

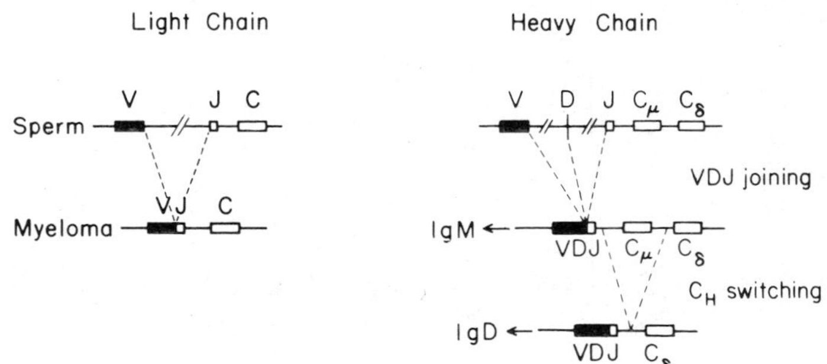

Fig. 1. Schematic diagram showing segments of a light and a heavy chain of an antibody molecule.

processed in four different ways to generate the four distinct polypeptides (Early et al., 1980b; Moore et al., 1981). One wonders whether this RNA splicing strategy may be employed in other informational molecules, such as Ia antigens, which appear to have alternative forms that must exist in very different environments, e.g., the membrane lipid bilay and the hydrophobic serum.

DNA rearrangements also may contribute in several ways to the generation of antibody diversity (Crews et al., 1981).

One wonders whether the strategies for information handling the antibody gene families that have evolved also might be employed in other complex eukaryotic systems requiring informationally sophisticated readouts. We believe that there will be a variety of complex area-code gene families which employ at least some of the genetic mechanisms used by the antibody gene families (Hood et al., 1977).

REFERENCES

Crews, S., J. Griffin, H. Huang, K. Calame, and L. Hood. (1981). *Cell*, 25, 59-66.

Davis, M. M., S. K. Kim, and L. Hood. (1980). *Cell*, 22, 1-2.

Early, P., H. Huang, M. Davis, K. Calame, and L. Hood. (1980a). *Cell*, 19, 981-992.

Early, P., J. Rogers, M. Davis, K. Calame, M. Bond, R. Wall, and L. Hood. (1980b). *Cell*, 20, 313-320.

Hood, L., H. Huang, and W. J. Dreyer. (1977). *J. Supramol. Struct.*, 7, 531-559.

Moore, K. W., J. Rogers, T. Hunkapiller, P. Early, C. Nottenburg, I. Weissman, H. Bazin, R. Wall, and L. E. Hood. (1981). *Proc. Nat. Acad. Sci., USA*, 78, 1800-1804.

STRUCTURE AND REGULATION OF DUCK GLOBIN GENES[1]

Gary V. Paddock*, Fu-Kuen Lin*, Robert Frankis**, William McNeill***, and Jim Gaubatz*

* Department of Basic and Clinical Immunology and Microbiology
** Molecular and Cellular Biology and Pathobiology Program
*** Infectious Diseases Division, Department of Medicine
Medical University of South Carolina
171 Ashley Avenue
Charleston, South Carolina 29425

[1]Publication No. 482 from the Department of Basic and Clinical Immunology and Microbiology, Medical University of South Carolina. Research supported in part by grants to G.P. from NIH (GM-24783) and from State of South Carolina Institutional Support Funds.

[2]Present address: Department of Biochemistry, MSB 2170, University of South Alabama, Mobile, Alabama 36688.

I. INTRODUCTION

Embryonic ducks have a changing complement of hemoglobins as they develop (Borgese and Bertles, 1965). HbIV is the earliest known hemoglobin, followed by HbIII (Table 1). HbIV disappears very early in development whereas HbIII is produced in gradually decreasing amounts until it is no longer present in adult ducks. Later work (Borgese and Nagel, 1977) shows that there are actually two embryonic hemoglobins that appear and disappear early in embryonic development. The adult hemoglobins, HbII and HbI, appear in the later embryonic stages with the relative amounts changing gradually to a final adult ratio of approximately 80% HbI to 20% HbII. When hemolytic anemia is induced in adult ducks by treatment with phenylhydrazine, the ratios of HbI and HbII change as shown in Table 1 (Bertles and Borgese, 1968). Based on gel electrophoresis of whole hemoglobins and limited aminto acid composition data (Saha and Ghosh, 1965), Bertles and Borgese speculated that the α^{I} and β^{I} globins were coordinately decreased while the α^{II} and β^{II} globins were coordinately increased. Low levels of HbIII and and unknown hemoglobin were also induced.

Table 1. Duck globins.

	Age (days)	HbIV	HbIII	HbII	HbI
	7	74%	26%	--	--
Embryonic	14	--	16%	49%	35%
	21	--	13%	44%	43%
(hatch)	28	--	--	--	--
	1	--	10%	33%	57%
Adult	14	--	8%	31%	61%
	84%	--	--	18%	82%

Anemic Shift	HbI		HbII	HbI : HbII	
	$\alpha_2^{I}\beta_2^{I}$	:	$\alpha_2^{II}\beta_2^{II}$	60%	40%
				+	
	80%		20%	small amounts of HbIII and Hbx	

Data abstracted from Borgese and Bertles (1965) and Bertles and Borgese (1968). In later work of Borgese and Nagel (1977), the following notation changes are made for the duck hemoglobins: IV=E1, III=F, II=D, I=A, and E2 designates a new embryonic hemoglobin.

At this point a brief review of the chicken globins is necessary, since the duck globin protein sequences were never determined and comparisons will be made throughout this discussion on both the expression and structure of duck and chicken globin genes. The chicken embryonic hemoglobins consist of HbP ($\pi_2\rho_2$), HbP' ($\pi_2\rho_2$), HbM ($\alpha_2^D\varepsilon_2$), and HbE ($\alpha_2^A\varepsilon_2$) (Brown and Ingram, 1974). There is also a distinct hatching hemoglobin, HbH ($\alpha_2^A\beta_2^H$) Villeponteau and Martinson, 1981; Dolan et al., 1981; Brown and Ingram, 1974). The adult hemoglobins are HbA ($\alpha_2^A\beta_2$) and HbD ($\alpha_2^D\beta_2$) (Brown and Ingram, 1974). Unlike in ducks, there appears to be only one β chain in chickens (Vandecasserie et al., 1975). The amino acid sequences are known for the embryonic alpha-like globins, π and π' (Chapman et al., 1980), embryonic beta-type globin ρ (Chapman et al., 1981), αA (Matsuda et al., 1971; Paul et al., 1974), α^D (Takei et al., 1975; Dodgson et al., 1981), and β (Matsuda et al., 1973; Richards et al., 1979; Vandecasserie et al., 1975). When adult chickens were made anemic through treatment with phenylhydrazine, a new set of α globins, α^s, were thought to be induced.[3] The amino acid sequences for these proteins have not been determined, but the mRNA sequences for several have been elucidated after prior conversion to recombinant cDNAs. Three of the sequences thus far determined are each slightly different from the other (Salser et al., 1979; Deacon et al., 1980; Richards and Wells, 1980). The α^s sequences of Salser et al. and Richards and Wells are quite close (2 amino acid codon differences) and could be alleles of the same gene. However, the sequence of Deacon et al. (1980) has additional amino acid codon differences at positions 92-94 which suggest either that it is coded for by a different gene or that their sequence is erroneous since all other alpha globins are invariant at these positions (Dayhoff, 1972).

Still different alpha globin genes have also been cloned in recombinant cDNA as determined by restriction enzyme analysis (Reynaud et al., 1980). Reynaud et al. suggest that these alpha

[3]Throughout this discussion to avoid confusion with previous literature, we refer to α^A for the normal adult avian alpha globin as determined by Matsuda et al. (1971). The anemic stress alpha globin of chickens is referred to as α^s, even though it might be the real α^A globin (Didgson et al., 1981), and α^a refers to the anemic stress alpha globin of ducks and is equivalent to α^s.

globin genes are for α^A and α^D. In contradiction, Richards and Wells (1980) suggested that the mRNA for α^A and α^D is not present in anemic chickens based on restriction enzyme digests of double stranded cDNA. It is possible, however, that the α^D cDNA was not synthesized in sufficient quantities to be seen in their gel autoradiographs. Dodgson et al. (1981) have shown that α^s may really be the normal adult α^A. Thus either the original α^A protein sequence of Matsuda et al. (1971) is drastically in error or, alternatively, α^A and α^s could be two different genes expressed in different strains of chickens. The data of Dodgson et al. (1981) also suggest that the α^D sequence of Takei et al. (1975) is incorrect or that as for α^A and α^s and there are strain differences. It thus appears that anemic chickens undergo a shift in globins (Stino and Washburn, 1970) which is similar although probably not identical to that observed for anemic ducks, and about which there is controversy as to its precise nature. The data of Stino and Washburn (1970), although showing changes in quantities of the two major hemoglobins when adult chickens are made anemic, imply that both α^A and α^D are present in both anemic and healthy adults. Their data also indicate that there are indeed strain differences in the chicken alpha D globins. We are currently characterizing the duck globin systems. By comparing the duck and chicken globin sequences and their expression, we hope to learn of the origin and regulation of the avian globin genes.

II. SEQUENCE ANALYSIS OF AN ANEMIC DUCK ALPHA GLOBIN GENE AND THE EVOLUTION OF AVIAN ALPHA GLOBINS

By the method of Maxam and Gilbert (1977, 1980), we have determined the nucleotide sequence (Fig. 1) of a recombinant cDNA for an alpha globin, α^a, from anemic ducks (Paddock and Gaubatz, 1981), which is closely related to goose α^A (Debouverie, 1975; Braunitzer and Oberthür, 1979) and chicken α^s and α^A globins but only distantly related to the chicken π and α^D globins (Table 2). We first compared the duck α^a sequence (Paddock and Gaubatz, 1981) with the first available chicken α^s sequence (Deacon et al., 1980) and found anomalies that suggested a possible dual origin for the gene. The duck α^a amino acid sequence was identical to goose and chicken α^A at positions 92-94 (Debouverie, 1975; Matsuda et al., 1971), and different from chicken α^s at all three positions, i.e., for the α^s of Deacon et al. (1980). One duck globin codon had three changes and another two changes from the codons for the anemic chicken alpha globin. The duck globin gene was more closely related to the anemic chicken α^s globin (Deacon et al., 1980) in the last one-third than in the first two-thirds of the protein coding region of the gene, i.e., 45 nucleotides different in the first 99 codons versus 12 nucleotides in the last 42 codons. Furthermore, while the chicken α^s amino acid sequence was less

1				5					10			13
Val	Leu	Ser	Ala	Ala	Asp	Lys	Thr	Asn	Val	Lys	Gly	Val
GUG ·	CUG ·	UCC ·	GCG ·	GCU ·	GAC ·	AAG ·	ACC ·	AAC ·	GUC ·	AAG ·	GGU ·	GUC ·
14	15					20					25	26
Phe	Ser	Lys	Ile	Gly	Gly	His	Ala	Glu	Glu	Tyr	Gly	Ala
UUC ·	UCC ·	AAA ·	AUC ·	GGU ·	GGC ·	CAU ·	GCU ·	GAA ·	GAG ·	UAU ·	GGC ·	GCC ·
27			30					35				39
Glu	Thr	Leu	Glu	Arg	Met	Phe	Ile	Ala	Tyr	Pro	Gln	Thr
GAG ·	ACC ·	CUG ·	GAG ·	AGG ·	AUG ·	UUC ·	AUC ·	GCC ·	UAC ·	CCC ·	CAG ·	ACC ·
40					45					50		52
Lys	Thr	Tyr	Phe	Pro	His	Phe	Asp	Leu	Ser	His	Gly	Ser
AAG ·	ACC ·	UAC ·	UUC ·	CCC ·	CAC ·	UUU ·	GAC ·	CUG ·	UCC ·	CAC ·	GGC ·	UCU ·
53		55					60					65
Ala	Gln	Ile	Lys	Ala	His	Gly	Lys	Lys	Val	Ala	Ala	Ala
GCU ·	CAA ·	AUC ·	AAG ·	GCC ·	CAU ·	GGC ·	AAG ·	AAG ·	GUG ·	GCG ·	GCU ·	GCC ·
66				70					75			78
Leu	Val	Glu	Ala	Val	Asn	His	Val	Asp	Asp	Ile	Ala	Gly
CUA ·	GUU ·	GAG ·	GCU ·	GUC ·	AAC ·	CAC ·	GUC ·	GAU ·	GAC ·	AUC ·	GCG ·	GGU ·
79	80					85					90	91
Ala	Leu	Ser	Lys	Leu	Ser	Asp	Leu	His	Ala	Gln	Lys	Leu
GCU ·	CUC ·	UCC ·	AAG ·	CUC ·	AGU ·	GAC ·	CUC ·	CAC ·	GCC ·	CAA ·	AAG ·	CUC ·
92			95					100				104
Arg	Val	Asp	Pro	Val	Asn	Phe	Lys	Phe	Leu	Gly	His	Cys
CGU ·	GUG ·	GAC ·	CCU ·	GUC ·	AAC ·	UUC ·	AAA ·	UUC ·	CUG ·	GGC ·	CAC ·	UGC ·
105					110					115		117
Phe	Leu	Val	Val	Val	Ala	Ile	His	His	Pro	Ala	Ala	Leu
UUC ·	CUG ·	GUG ·	GUG ·	GUU ·	GCC ·	AUC ·	CAC ·	CAC ·	CCU ·	GCU ·	GCC ·	CUG ·
118		120					125					130
Thr	Pro	Glu	Val	His	Ala	Ser	Leu	Asp	Lys	Phe	Met	Cys
ACC ·	CCA ·	GAG ·	GUC ·	CAC ·	GCU ·	UCC ·	CUG ·	GAC ·	AAG ·	UUC ·	AUG ·	UGC ·
131				135					140	141		
Ala	Val	Gly	Ala	Val	Leu	Thr	Ala	Lys	Tyr	Arg		
GCC ·	GUG ·	GGU ·	GCU ·	GUG ·	CUG ·	ACU ·	GCC ·	AAG ·	UAC ·	CGU ·		

Term
UAG · ACGGCACCGUGGCUAGAGCUGGACCCACCCUGUUGCCAGCCUUCCAACUGCGAGCAGCCAAAUGAUCUGAAAU
AAAAUCUGUUGCAUUUGUGCUCC poly(A)

Figure 1. Sequence of gene for duck α^a globin. Predicted duck amino acids are above the line of nucleotides. Nucleotides which differ in sequence from that for the alpha globin gene sequence obtained for anemic chickens are underlined (Richards and Wells, 1980). Nucleotides in the chicken sequence missing in duck are indicated by (∧); the reverse situation is indicated by (∨).

Table 2. Comparison of amino acid differences in avian alpha globins.

	Number of Differences						
	Duck[1] Alpha a	Chicken[2] Alpha s	Ostrich[3] Alpha A	Goose[4] Alpha A	Chicken[5] π	Chicken[6] Alpha D	Duck[7] Alpha D
Duck Alpha a		18	17	7	61	61	52
Chicken Alpha s	13		18	15	61	59	58
Ostrich Alpha A	12	13		15	60	60	57
Goose Alpha A	5	11	11		60	62	59
Chicken π	43	43	43	43		58	63
Chicken Alpha D	43	42	42	44	41		27
Duck Alpha D	37	41	40	42	45	18	
	Percent Differences						

[1]Paddock and Gaubatz, 1981; [2]Richards and Wells, 1981; [3]Oberthür et al., 1980; [4]Braunitzer and Oberthür, 1979; [5]Chapman et al., 1980; [6]Dodgson et al., 1981; [7]Preliminary sequence, this paper.

related to chicken α^A for codons 1-99 than for codons 100-141, the duck α^a was more related to chicken and goose α^A and less related to chicken α^s for codons 1-99 than for codons 100-141 (Table 3). While the amino acids at positions 92-94 were invariant for adult alpha globins (Dayhoff, 1972), we considered the

Table 3. Percent amino acid differences.

	[2]Chicken Alpha s	[3]Chicken Alpha s	[4]Chicken Alpha A	[5]Goose Alpha A
[1]Duck Alpha a				
Amino acids 1-99	17%	14%	17%	14%
Amino acids 100-141	12%		26%	33%
[2]Chicken Alpha s				
Amino acids 1-99	--	--	18%	20%
Amino acids 100-141	--	--	14%	21%
[3]Chicken Alpha s				
Amino acids 1-99	--	--	15%	17%
Amino acids 100-141	--	--	17%	24%

[1]Paddock and Gaubatz, 1981; [2]Deacon et al., 1980; [3]Richards and Wells, 1981; [4]Matsuda et al., 1971; [5]Debouverie, 1975.

possibility of variance at those positions for this different chicken alpha globin. We thus further considered the possibility that the duck α^a gene might have evolved in parts from more than one ancestor alpha globin gene (Paddock and Gaubatz, 1981). Possibilities included a Lepore-like fusion (Benz and Forget, 1975) or a different series of RNA splicing steps to produce the mRNA product seen during anemia. Admittedly, our division of the mRNA molecule at the codons for amion acids 99 and 100 was arbitrarily chosen to correspond with the intron-exon junction found there for alpha globins (Leder et al., 1978), but this seemed fair given the nearby amino acid codon differences at positions 92-94. Membrane and serum IgM heavy chains are documented examples of different protein sequences produced at the RNA splicing step of mRNA processing in eukaryotes (Alt et al., 1980; Rogers et al.,

1980; Early et al., 1980). However, others (Salser et al., 1979; Richards and Wells, 1980; Dodgson et al., 1981) have shown that the sequence at positions 92-94 for the chicken α^s globin are the same as for all other alpha globins including the duck α^a globin. Chapman et al. (1980) have shown that the embryonic alpha-like π globin in chickens also have the invariant amino acid sequence at positions 92-94. Furthermore, Dodgson et al. (1981) have challenged the existence of a gene for the α^A of Matsuda et al. (1971) and claim that the α^s globin is the correct adult α^A globin (there are 22 amino acid codon differences for α^s and α^A). Their hypothesis is based on nucleotide sequence analysis of recombinant DNAs contaïning cDNA or genome fragments. In addition, Braunitzer and Oberthür (1979) have challenged the goose sequence of Debouverie (1975) (there are 27 amino acid differences in the two sequences). Interestingly enough, Braunitzer and Oberthür's goose sequence (amino acid sequence for α^A globin from nonanemic geese) agrees far more closely with the nucleic acid sequences determined for the chicken α^s and duck α^a globin genes than with the protein sequence determined for chicken α^A. On the other hand, Debouverie's (1975) goose sequence agrees most closely with that of Matsuda et al. (1971) for chicken.

At this point it would seem we are left with two possibilities. The first possibility is that Matsuda et al. (1971) were incredibly erroneous in their sequence determination of the chicken α^A globin and that Debouverie (1975) followed suit for the goose α^A globin. But we must remember that the Matsuda group was able to determine the beta globin amino acid sequence perfectly (Matsuda et al., 1973 vs. Richards et al., 1979). Another possibility is strain differences among the different avian species. Such differences found by Oberthür et al. (1980) for the α^A globins of domestic and wild geese. As an alternative we propose that a gene duplication of α^A globins occurred in avians followed by substantial divergence. Different avian species may have handled this availability of two α^A genes (one of which corresponds to α^s) in different ways. Chickens and geese may have evolved with one gene or the other being silenced in different subpopulations, as Stino and Washburn (1970) have observed for the alpha D globin in different strains of chickens. Perhaps both α^A and α^s are active in some avian species. Ducks may have evolved similarly or perhaps evolved with a fusion of the originally split genes. When the duck α^a sequence is compared with that of Richards and Wells (1980) instead of Deacon et al. (1980) for chicken α^s (Table 3), the evidence for a dual origin is less impressive but still substantial. When one compares duck α^a with chicken or goose α^A, the relative differences in homology for the first two-thrids versus the last one-third of the amino acid codons are still far greater than for either chicken α^s sequence.

Therwath et al. (1980) and Reynaud et al. (1980) have previously claimed to have isolated alpha A globin recombinant cDNAs from anemic chickens and ducks based on restriction enzyme cleavage data. We (Paddock and Gaubatz, 1981) had disputed that because of the suggestion by Richards and Wells (1980) that alpha A globins were not found in anemic chickens. In light of our "dual gene evolution hypothesis" for avian alpha globins, we now suggest that Therwath et al. (1980) and Reynaud et al. (1980) may indeed have found the long sought α^A genes. If indeed they have, then their findings support the dual gene hypothesis and also show that either α^s or α^a can be widespread, as opposed to the hypothesis of Dodgson et al. (1981) that chickens with the α^A gene would be found only in nonwestern countries such as Japan. We have placed the α^a globin into position with other avian globins in an evolutionary family tree (Fig. 2). It appears that the α^a globins are a uniquely avian feature in that their divergence from the α^A globins approximately 122 to 130 million years ago corresponds quite closely with the emergence of modern birds, estimated to be 135 million years ago (Dodson, 1960). With the chicken and goose amino acid sequences previously available for comparison, we (Paddock and Gaubatz, 1981) had observed that strangely enough the alpha A globins of geese and chicken appeared to be as closely or even more closely related than those of geese and ducks. As can be seen from Table 2 when the analogous globins are compared, however, the alpha globins of geese and ducks are far more closely related to each other than either is to chicken alpha globin, as one would expect from the morphology of the animals.

An interesting evolutionary comparison can be made with the geese alpha globins. Oberthür et al. (1980) have found that a wild goose strain has Ser, Val, and Ala at positions 18, 63, and 119, respectively, while domestic geese have Gly, Ala, and Pro (Braunitzer and Oberthür, 1979). We have found that domestic ducks have the same sequence at those positions (Paddock and Gaubatz, 1981) as the domestic geese. Admittedly we do not yet have the corresponding sequence for a wild strain of ducks, but assuming that ducks and geese evolved as separate species first in the wild, then it would appear that a subpopulation in each species underwent a parallel set of identical mutations in their alpha globin genes while concurrently becoming ground swellers. The ground dwellers then eventually were domesticated.

III. SEQUENCE ANALYSIS OF DUCK ALPHA D GLOBIN

We have also determined a preliminary sequence for duck α^D globin (Fig. 3). As can be seen from Table 2, duck α^D globin and about equidistant from the α^A, α^s, or α^a globins and the π globins. The evolutionary distance between the two avian α^D globins (18% codon differences) is what would be expected for

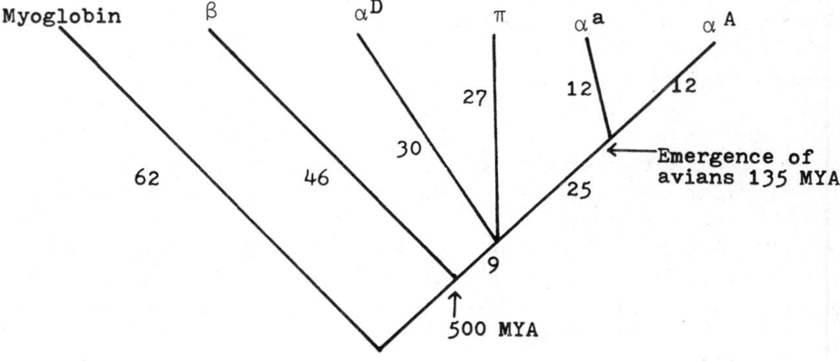

Avian Globins

Fig. 2. Evolutionary family tree for avian globins (Chapman et al., 1980; Paddock and Gaubatz, 1981; Matsuda et al., 1971; Deacon et al. 1980). The divergence of the alpha and beta globins occurred about 500 million years ago (MYA) and the emergence of modern birds about 135 million years ago (Dodson, 1960). The divergence of α^a (α^s) and α^A is thus estimated to have occurred about 130 million years ago. If the data of Richards and Wells (1980) are used for α^s, then the divergence of α^a (α^s) from α^A would show 11 base substitutions in each branch, and the time of divergence would be about 122 million years ago.

									11		13	
Leu	Thr	Ala	Glu	Asp	Lys	Lys	Leu	Ile	Thr	Gln	Leu	Trp
CUG •	ACC •	GCC •	GAG •	GAC •	AAG •	AAG •	CUC •	AUC •	ACG •	CAG •	UUG •	UGG •
		17		19							26	
Glu	Lys	Val	Ala	Gly	His	Gln	Glu	Glu	Phe	Gly	Ser	Glu
GAG •	AAG •	GUG •	GCU •	GGC •	CAC •	CAG •	GAG •	GAA •	UUC •	GGA •	AGU •	GAA •
		30				34	35					
Ala	Leu	Gln	Arg	Met	Phe	Leu	Ala	Tyr	Pro	Gln	Thr	Lys
GCU •	CUG •	CAG •	AGG •	AUG •	UUC •	CUC •	GCC •	UAC •	CCC •	CAG •	ACC •	AAG •
								49				53
Thr	Tyr	Phe	Pro	His	Phe	Asp	Leu	His	Pro	Gly	Ser	Glu
ACC •	UAC •	UUC •	CCC •	CAC •	UUC •	GAC •	CUG •	CAU •	CCC •	GGC •	UCU •	GAA •
									63	64		
Gln	Val	Arg	Gly	His	Gly	Lys	Lys	Val	Ala	Ala	Ala	Leu
CAG •	GUC •	CGU •	GGC •	CAU •	GGC •	AAG •	AAA •	GUG •	GCA •	GCU •	GCC •	CUG •
					72	73						
Gly	Asn	Ala	Val	Lys	Ser	Leu	Asp	Asn	Leu	Ser	Gln	Ala
GGC •	AAU •	GCC •	GUG •	AAG •	AGC •	CUG •	GAC •	AAC •	CUC •	AGC •	CAG •	GCC •
80	81											
Leu	Ser	Glu	Leu	Ser	Asn	Leu	His	Ala	Tyr	Asn	Leu	Arg
CUG •	UCU •	GAG •	CUC •	AGC •	AAC •	CUG •	CAC •	GCC •	UAC •	AAC •	CUA •	CGU •
											G	
									102			105
Val	Asp	Pro	Val	Asn	Phe	Lys	Leu	Leu	Ala	Gln	Cys	Phe
GUU •	GAC •	CCC •	GUC •	AAC •	UUC •	AAG •	CUG •	CUG •	GCG •	CAG •	UGC •	UUC •
					U							
	107		109	110	111		113					118
Gln	Val	Val	Leu	Ala	Ala	His	Leu	Gly	Lys	Asp	Tyr	Ser
CAG •	GUG •	GUG •	CUG •	GCC •	GCA •	CAC •	CUG •	GGC •	AAA •	GAC •	UAC •	AGC •
		121								129		
Pro	Glu	Met	His	Ala	Ala	Phe	Asp	Lys	Phe	Met	Ser	Ala
CCC •	GAG •	AUG •	CAU •	GCU •	GCC •	UUU •	GAC •	AAG •	UUC •	AUG •	UCC •	GCC •
	133											
Val	Ala	Ala	Val	Leu	Ala	Glu	Lys	Tyr	Arg	Term		
GUG •	GCU •	GCC •	GUG •	CUG •	GCU •	GAA •	AAG •	UAC •	AGA •	UGA •		

GCCACUGCCUGCACCCUUGCACCUUCAAUAAAGACACCAUUACCAC poly(A)

Fig. 3. Preliminary sequence for duck α^D globin gene from the codon for amino acid 2 through 3' untranslated region. Predicted duck amino acids are above the line of nucleotides. The numbered amino acids signify amino acids which are different from that of chicken α^D globin (Dodgson et al., 1981). One of our recombinants contained CUA and UUC for condons 91 and 98, respectively, while another recombinant had CUG and UUU for those positions.

alpha globins of similar function, i.e., for chicken α^s and duck α^a the difference is 13%. We note here a most interesting feature of this gene. The 3' untranslated region is only about one-half of the length observed for other alpha globin genes. Thus we are able to determine the essence of what is important for 3' end sequences of globin genes. When the α^D 3' end is compared with the α^a 3' end, there is very little homology (Fig. 4). This is in contrast to the 85% homology observed for the chicken α^s and duck α^a sequences shown in Figure 1. Figure 4 shows several regions of homology: I (PuCCCU), found in all alpha globin mRNA sequenced to date except human alpha-1 globin (Michelson and Orkin, 1980); II (AAUAAA), found in poly(A)-tailed mRNA; III (CAUPy), found in all avian globin mRNA (including beta globin); and IV (C), found in all globin mRNA except *Xenopus* beta globin. Nucleotides 6-8 (CAC) also appear in all avian globin genes and nucleotides 12-14 (CUG) appear in avian alpha globin genes. We hypothesize that these regions are important to the proper functioning of the mRNA for these respective gene classes either through their effect on mRNA structure or as binding sites for porteins required for mRNA function or stability.

Considering the cross-species conservation of duck α^a and chicken α^s 3' end sequences, a converse argument indicates that these 3' end sequences are very important for the regulation and expression of the particular gene concerned (i.e., when the globin mRNA is synthesized and in what quantity). It has been noted that proteins with similarity in function are more closely related in different species than are proteins which evolved to slightly different function from the same ancestor genes within a species (Melderis et al., 1974; Chapman et al., 1980). For example, the embryonic alpha-like globins between species are more closely related than the embryonic alpha and adult alpha globins within the same species. Thus, just as a protein must carry out its function in its own peculiar fashion, so must the mRNA be synthesized and be available to act as a template for the synthesis of its respective protein at the right time and in the correct amount.

IV. DUCK GLOBINS AND THE ANEMIC SHIFT

Because of the discovery of the anemic alpha globins in both chickens and ducks, we thought the anemic shift in these avians might be more complicated than first hypothesized by Bertles and Borgese (1968). It appears that α^D globin is present in anemic ducks, while in anemic chickens it is controversial (Richards and Wells, 1980; Reynaud et al., 1980) but probable (Stino and Washburn, 1970). This is probably the minor α^{II} globin of Borgese and Bertles and thus must increase when the animals are made anemic. Because of the paper by Dodgson et al. (1981), we cannot be certain whether the major alpha globin species, α^I, merely

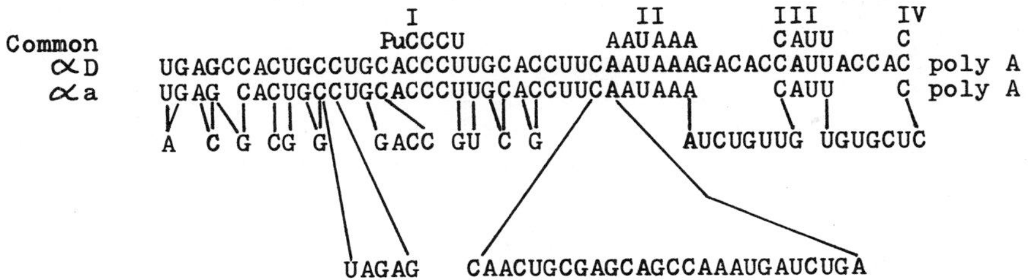

I All Alpha Globins
II All mRNA (with poly A)
III All Avian Globins
IV All Globins Except Xenopus Beta Globins

Fig. 4. Comparison of the 3' untranslated regions for the duck α^D and α^a globin genes. Sequences common to other genes are also shown (Salser et al., 1979; Richards et al., 1979; Richards and Wells, 1980; Williams et al., 1980; Proudfoot and Maniatis, 1980; Heindell et al., 1978). Note that the common region, AAUAAA, is modified in some other poly(A)-terminated mRNAs. For example, AUUAAA has recently been found for mouse pancreatic amylase mRNA (Hagenbuchle et al., 1980).

decreases or whether an entirely different alpha globin is substituted when the ducks are made anemic. There are still different duck alpha globin genes isolated by others (Therwath et al., 1980; Reynaud et al., 1980) from anemic ducks and chickens, which have different restriction enzyme sites than those sequenced thus far. Thus the catalog of globin genes expressed in anemic and non-anemic ducks is becoming quite complex. Our dual gene evolution hypothesis would predict that α^a or α^A can be expressed in either healthy or anemic ducks (α^S or α^A chickens) depending on which is normally expressed in the particular strain used and that no exchange of alpha globins occurs. Isolation of the globin proteins from their respective hemoglobins and subsequent peptide mapping or sequence analysis of these proteins will be required to demonstrate the identity of the globins synthesized in non-anemic and anemic ducks, especially since even the existence of two rather than one beta globin for adult ducks is not proven and in fact has been disputed (Spohr et al., 1972).

V. ORGANIZATION OF BETA GLOBIN GENES

We have constructed a gene library of duck genome fragments (from a partial Eco RI digestion) contained in lambda bacteriophage ($\lambda gt_{WES} \cdot \lambda^B$) recombinant DNAs. We determined which of these contained globin genes by hybridization to ^{125}I-labeled globin mRNA in which the poly(A) tails were removed with RNase H. Three of these recombinants were found to hybridize to a β globin recombinant cDNA but not to α^a or α^D recombinant cDNAs. In addition, each of the three recombinants showed the large second intron loop characteristic of beta globin genes when they were subjected to R-loop analysis and visualized by electron microscopy (Fig. 5) as in Kaback et al. (1979). Two of these recombinants have been partially mapped (Fig. 6) using Southern's (1975, 1979) gel technique. We still must subclone smaller fragments from these recombinants into pBR322 in order to obtain a finer analysis. However, our results indicate that each recombinant contains at least two closely packed β-type globin genes, since the size of a β-type globin gene is approximately 1500-1800 base pairs including introns, whereas the regions which hybridize to the ^{125}I-labeled mRNA probe are much larger. A number of experiments, both *in vitro* and *in vivo* (Grosveld et al., 1981; Grosschedle and Birnstiel, 1980; Dierks et al., 1981) have shown that regions such as the Goldberg-Hogness box (Gannon et al., 1979) just upstream of the capping box are responsible for initiating transcription of pre-mRNA which is then processed and modified to mature mRNA. In view of the close packing we have found for the duck globin genes and others have found for the chicken globin genes, and the large 27S pre-mRNA species found by Niessing and Scherrer (Imaizumi et al., 1973; Spohr et al., 1972; Reynaud et al., 1980; Niessing, 1978), it seems likely that additional secondary promotor sites exist further upstream for transcription of these very large pre-mRNA species. Any biological role for this second promoter - large pre-mRNA system must remain speculatory at this time, however, since there is no proof that they are processed to mature mRNA as are the smaller 15S-18S pre-mRNA species.

When we obtain nucleotide sequences for these genes, they will be compared with the sequences for chicken β-type globins and with our own duck β globin recombinant cDNA nucleotide sequences (Fig. 7) and those of Hampe et al. (1981). Such a comparison might uncover any special feature in the genome sequences which could account for the extremely high degree of preservation observed when the duck and chicken β globin mRNA sequences are compared (2.8% amino acid changes and 5.6% nucleotide changes for codons 76 through termination). As can be seen from Table 2, this is a far greater degree of conservation than is found for the alpha globins. Conservation of avian globin genes, including chicken and duck, has also been noted by Engel and Dodgson (1978).

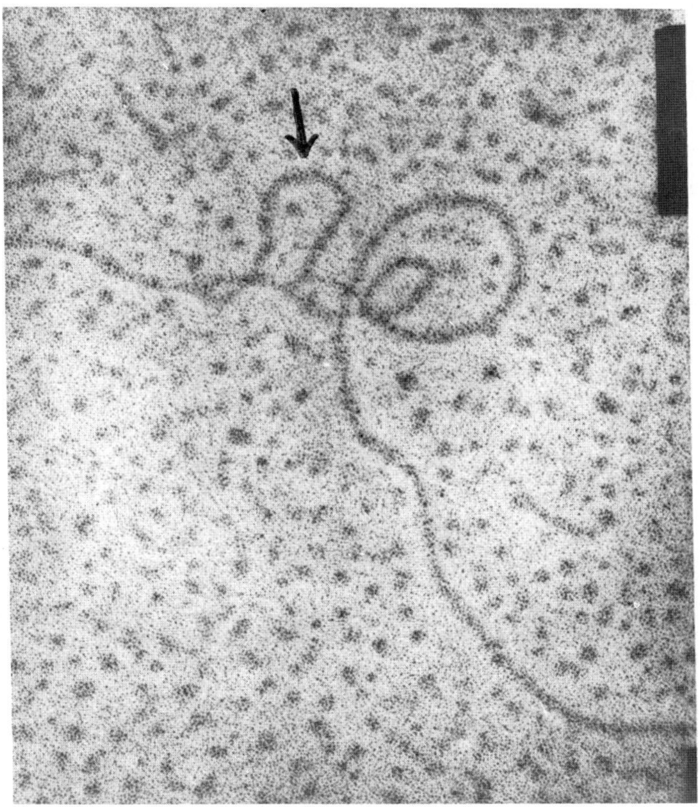

Fig. 5. R-loop analysis of λDGP-1. Hybridization of duck globin mRNA to λDGP-1 DNA is shown by electron microscopy to have the large loop (arrow) characteristic of the second intron in a beta globin gene. The two adjacent exon loops are also seen.

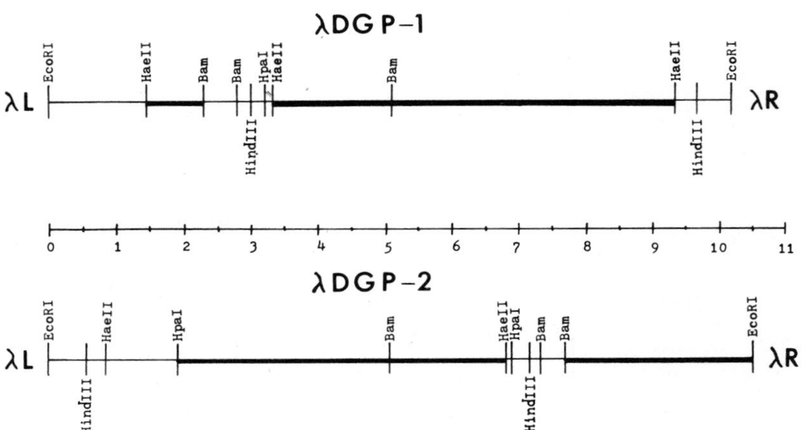

Fig. 6. Preliminary map of two recombinant lambda bacteriophage containing beta-type globin genes. The map was determined using the technique of Southern (1975, 1979).

Our more recent mapping of the genome recombinants λDGP-1 and λ DGP-2 (F. K. Lin and G. Paddock, unpubl.) has further refined the maps depicted in Figure 6. Numerous restriction enzyme sites have been conserved in the two gene pairs, indicating a fairly recent gene duplication event. Conserved sites appear both within and adjacent to the coding regions as well as in regions quite far removed from the coding sequences. It is in these extragenic conserved sites (found from both interspecies and intraspecies comparisons) that we will search for possible promoter and other regulatory sequences.

VI. ACKNOWLEDGEMENTS

We gratefully acknowledge technical assistance by Louise Just and Conrad Lau, manuscript preparation by Linda Paddock and Nancy Butler, and editorial assistance by Charles Smith.

```
                                                                  76    77
                                                                        Asn
                                                                  XAG · AAC ·

            80                            85    *                       90
Leu   Asp   Asn   Ile   Lys   Asn   Thr   Phe   Ala   Gln   Leu   Ser   Glu
CU()· GAC · AAC · AUC · AAG · AAC · ACC · UUC · GCC · CAG · CUG · UCC · GAG ·
                                                X     X                 X

                        95                            100
Leu   His   Cys   Asp   Lys   Leu   His   Val   Asp   Pro   Glu   Asn   Phe
CUG · CAC · UGC · GAC · AAG · CUG · CAC · GUG · GAC · CCC · GAG · AAC · UUC ·
      X     X                       X

      105                           110                           115
Arg   Leu   Leu   Gly   Asp   Ile   Leu   Ile   Ile   Val   Leu   Ala   Ala
AGG · CUC · CUG · GGU · GAC · AUC · CUC · AUC · AUC · GUC · CUG · GCC · GCC ·
                                                X

            *     120                           125
His   Phe   Thr   Lys   Asp   Phe   Thr   Pro   Glu   Cys   Gln   Ala   Ala
CAC · UUC · ACC · AAG · GAU · UUC · ACU · CCU · GAA · UGC · CAG · GCU · GCC ·
            X           X

130                           135                           140
Trp   Gln   Lys   Leu   Val   Arg   Val   Val   Ala   His   Ala   Leu   Ala
UGG · CAG · AAG · CUG · GUC · CGC · GUG · GUG · GCC · CAC · GCU · CUG · GCC ·
                                                      X     X           X

            145
Arg   Lys   Tyr   His   Term
CGC · AAG · UAC · CAC · UAA ·
```

Fig. 7. Preliminary sequence of gene for duck β globin from the codon for amino acid 76 through the termination condon. Predicted amino acids are above the line of nucleotides. The asterisk (*) signifies an amino acid and and (X) signifies a nucleotide which is different from that of chicken globin (Richards et al., 1979). The sequence is in agreement with the recently published sequence of Hampe et al. (1981).

VII. REFERENCES

Alt, F. W., Bothwell, A. L. M., Knapp, M., Siden, E., Mather, E., Koshland, M., and Baltimore, D. (1980). Synthesis of Secreted and Membrane-Bound Immunoglobulin Mu Heavy Chains is Directed by mRNAs that Differ at their 3' Ends. *Cell*, 20, 293-301.

Benz, E. J., Jr. and Forget, B. G. (1975). The Molecular Genetics of the Thalassemia Syndromes. *Progress in Hematology*, 9, 107-155.

Bertles, J. F. and Borgese, T. A. (1968). Disproportional Synthesis of the Adult Ducks Two Hemoglobins During Acute Anemia. *J. Clin. Invest.*, 47, 679-689.

Borgese, T. A. and Bertles, J. F. (1965). Hemoglobin Heterogeneity: Embryonic Hemoglobin in the Duckling and its Disappearance in the Adult. *Science*, 148, 509-511.

Borgese, T. A. and Nagel, R. L. (1977). Differential Effects of 2,3-DPG ATP and Inositol Pentaphosphate (IP5) on the Oxygen Equilibria of Duck Embryonic, Fetal and Adult Hemoglobins. *Comp. Biochem. Physiol.*, 56A, 539-543.

Braunitzer, G. and Oberthür, W. (1979). Die Primärstruktur des Hämoglobins der Graugans (*Anser anser*) Die ungleiche Evolution der Beta-Ketten (Versuch einer biochemischen Analyse des Verhaltens). Hoppe-Seyler's *Z. Physiol. Chem.*, 360, 679-683.

Brown, J. L. and Ingram, V. M. (1974). Structural Studies on Chick Embryonic Hemoglobins. *J. Biol. Chem.*, 249, 3960-3972.

Chapman, B. S., Tobin, A. J., and Hood, L. E. (1980). Complete Amino Acid Sequences of the Major Early Embryonic Alpha-Like Globins of the Chicken. *J. Biol. Chem.*, 255, 9051-9059.

Chapman, B. S., Tobin, A. J., and Hood, L. E. (1981). Complete Amino Acid Sequence of the Major Early Embryonic Beta-like Globin in Chicken. *J. Biol. Chem.*, 256, 5524-5531.

Dayhoff, M. O. (1972). *Atlas of Protein Sequence and Structure*, vol. 5.

Deacon, N. J., Shine, J., and Naora, H. (1980). Complete Nucleotide Sequence of a Cloned Alpha Globin cDNA. *Nucl. Acids Res.*, 8, 1187-1199.

Debouverie, D. (1975). Structure Primaire de la Chaine Alpha du Composant Majeur de Phémoglobine d'Oie (*Anser anser*). *Biochimie*, 57, 569-578.

Dierks, P., van Ooyen, A., Mantei, N., and Weissman, C. (1981). DNA Sequences Preceding the Rabbit Beta-Globin Gene Are Required for Formation in Mouse L Cells of Beta-Globin RNA with the Correct 5' Terminus. *Proc. Natl. Acad. Sci. USA*, 78, 1411-1415.

Dodgson, J. B., McCune, K. C., Rusling, D. J., Krust, A., and Engel, J. D. (1981). Adult Chicken Alpha-Globin Genes, Alpha A and Alpha D: No Anemic Shock Alpha-Globin Exists in Domestic Chickens. *Proc. Natl. Acad. Sci. USA*, 78, 5998-6002.

Dodson, E. O. (1960). "Evolution: Process and Product." Revised ed., Reinhold, N.Y.

Dolan, M., Sugarman, B. J., Dodgson, J. B., and Engel, J. D. (1981). Chromosomal Arrangement of the Chicken Beta-Type Globin Genes. *Cell*, 24, 669-677.

Early, P., Rogers, J., Davis, M., Calame, K., Bond, M., Wall, R., and Hood, L. (1980). Two mRNAs can be Produced from a Single Immunoglobulin Mu Gene by Alternative RNA Processing Pathways. *Cell*, 20, 313-319.

Engel, J. D. and Dodgson, J. B. (1978). Analysis of the Adult and Embryonic Chicken Globin Genes in Chromosomal DNA. *J. Biol. Chem.*, 253, 8239-8246.

Gannon, F., O'Hare, K., Perrin, F., LePennec, J. P., Benoist, C., Cochet, M., Breathnach, R., Royal, A., Garapin, A., Cami, B., and Chambon, P. (1979). Organization and Sequences at the 5' End of a Cloned Complete Ovalbumin Gene. *Nature*, 278, 428-434.

Grosschedle, R. and Birnstiel, M. L. (1980). Identification of Regulatory Sequences in the Prelude Sequences of an H2A Histone Gene by the Study of Specific Deletion Mutants *in vivo*. *Proc. Natl. Acad. Sci. USA*, 77, 1432-1436.

Grosveld, G. C., Shewmaker, C. K., Jat, P., and Flavell, R. A. (1981). Localization of DNA Sequences Necessary for Transcription of the Rabbit Beta-Globin Gene *in vitro*. *Cell*, 25, 215-226.

Hagenbuchle, O., Bovey, R., and Young, R. A. (1980). Tissue Specific Expression of Mouse Alpha Amylase Genes: Nucleotide Sequence of Isoenzyme mRNAs from Pancreas and Salivary Glands. *Cell*, 21, 179-187.

Hampe, A. Therwath, A., Soriano, P., and Galibert, F. (1981). Nucleotide Sequence Analysis of a Cloned Duck Beta-Globin cDNA. *Gene*, 14, 11-21.

Heindell, H. C., Liu, A., Paddock, G. V., Studnicka, G. M., and Salser, W. A. (1978). The Primary Sequence of Rabbit Alpha-Globin mRNA. *Cell*, 15, 43-54.

Imaizumi, T., Diggelmann, H., and Scherrer, K. (1973). Demonstration of Globin Messenger Sequences in Giant Nuclear Precursors of Messenger RNA of Avian Erythroblasts. *Proc. Natl. Acad. Sci. USA*, 70, 1122-1126.

Kaback, D. B., Angerer, L. M., and Davidson, N. (1979). Improved Methods for the Formation and Stabilization of R-loops. *Nucl. Acids Res.*, 6, 2499-2517.

Leder, A., Miller, H. I., Hamer, D. H., Seidman, J. G., Norman, B., Sullivan, M., and Leder, P. (1978). Comparison of Cloned Mouse Alpha- and Beta-Globin Genes: Conservation of Intervening Sequence Locations and Extragenic Homology. *Proc. Natl. Acad. Sci. USA*, 75, 6187-6191.

Matsuda, G., Takei, H., Wu, K. C., and Shiozawa, T. (1971). The Primary Structure of the Alpha Polypeptide Chain of AII Component of Adult Chicken Hemoglobin. *Int. J. Prot. Res.*, 3, 173-174.

Matsuda, G., Maita, T., Mizuno, K., and Ota, H. (1973). Amino Acid Sequence of a Beta Chain of AII Component of Adult Chicken Haemoglobin. *Nature New Biol.*, 244, 244.

Maxam, A. M. and Gilbert, W. (1977). A New Method for Sequencing DNA. *Proc. Natl. Acad. Sci. USA*, 74, 560-564.

Maxam, A. M. and Gilbert, W. (1980). Sequencing End-Labeled DNA with Base-Specific Chemical Cleavages. *Methods Enzymol.*, 65, 499-560.

Melderis, J., Steinheider, G., and Ostertag, W. (1974). Evidence for a Unique Kind of Alpha-Type Globin Chain in Early Mammalian Embryos. *Nature*, 250, 774-776.

Michelson, A. M. and Orkin, S. H. (1980). The 3' Ultranslated Regions of the Duplicated Human Alpha-Globin Genes Are Unexpectedly Divergent. *Cell*, 22, 371-377.

Niessing, J. (1978). Globin Messenger Precursor RNA in Duck Immature Red Blood Cells. *Eur. J. Biochem.*, 91, 587-598.

Oberthür, W., Voelter, W., and Braunitzer, G. (1980). Die Sequenz der Hämoglobine von Streifengans (*Anser indicus*) and Strauss (*Struthio camelus*). Inositpentaphosphat als Modulator der Evolutionsqeschwindigkeit: Die überraschende Sequenz Alpha 63 (E12) Valin. *Hoppe-Seyler's Z. Physiol. Chem.*, 361, 969-975.

Paddock, G. V. and Gaubatz, J. (1981). Nucleotide Sequence for a Novel Duck Alpha Globin Gene. *Eur. J. Biochem.*, 117, 269-273.

Paul, C., Vandecasserie, C., Schnek, A. G., and Leonis, J. (1974). N-Terminal Amino Acid Sequences of the Alpha and Beta Chains of the Two Chicken Hemoglobin Components. *Biochim. Biophys. Acta*, 371, 155-158.

Proudfoot, N. J. and Maniatis, T. (1980). The structure of a Human Alpha-Globin Pseudogene and its Relationship to Alpha-Globin Gene Duplication. *Cell*, 21, 537-544.

Reynaud, C. -A., Tahar, S. B., Krust, A., Franco, M. -P. A. de L., Goldenberg, S., Gannon, F., and Scherrer, K. (1980). Restriction Mapping of cDNA Recombinants Including the Adult Chicken and Duck Globin Messenger Sequences: A Comparative Study. *Gene*, 11, 259-269.

Richards, R. I., Shine, J., Ulbrich, A., Wells, J. R. E., and Goodman, H. M. (1979). Molecular Cloning and Sequence Analysis of Adult Chicken Beta Globin cDNA. *Nucl. Acids Res.*, 7, 1137-1146.

Richards, R. I. and Wells, J. R. E. (1980). Chicken Clobin Genes: Nucleotide Sequence of cDNA Clones Coding for the Alpha-Globin Expressed during Hemolytic Anemia. *J. Biol. Chem.*, 255, 9306-9311.

Rogers, J., Early, P., Carter, C., Calame, K., Bond, M., Hood, L., and Wall, R. (1980). Two mRNAs with Different 3' Ends Encode Membrane-Bound and Secreted Forms of Immunoglobulin Mu Chain. *Cell*, 20, 303-312.

Saha, A. and Ghosh, J. (1965). Comparative Studies on Avian Hemoglobins. *Comp. Biochem. Physiol.*, 15, 217-235.

Salser, W. A., Cummings, I., Liu, A., Strommer, J., Padayatty, J., and Clarke, P. (1979). Analysis of Chicken Globin cDNA Clones: Discovery of a Novel Chicken Alpha-Globin Gene Induced by Stress in Young Chickens. In "Cellular and Molecular Regulation of Hemoglobin Switching." (G. Stamatoyannopoulos and A. W. Nienhius, ed.) pp. 621-643, Grune and Stratton, N.Y.

Southern, E. M. (1975). Detection of Specific Sequences Among DNA Fragments Separated by Gel Electrophoresis. *J. Mol. Biol.*, 98, 503-517.

Southern, E. (1979). Gel Electrophoresis of Restriction Fragments. *Methods Enzymol.*, 68, 152-176.

Spohr, G., Imaizumi, T., Stewart, A., and Scherrer, K. (1972). Identification of Free Cytoplasmic Globin mRNA of Duck Erythroblasts by Hybridization to Anti-Messenger DNA and by Cell-Free Protein Synthesis. *FEBS Letters*, 28, 165-168.

Stino, F. K. R. and Washburn, K. W. (1970). Response of Chickens with Different Hemoglobin Genotypes to Phenylhydrazine-induced Anemia. 1. Electrophoretic Properties of Hemoglobin. *Poult. Sci.*, 49, 101-114.

Takei, H., Ota, Y., Wu, K. C., Kiyohara, T., and Matsuda, G. (1975). Amino Acid Sequence of the Alpha Chain of Chicken AI Hemoglobin. *J. Biochem.*, 77, 1345-1347.

Therwath, A., Soriano, P., and Scherrer, K. (1980). Analysis of Adult Duck Alpha and Beta Globin c-DNA Recombinant Plasmids. *Biochem. Int.*, 1, 32-40.

Vandecasserie, C., Paul, C., Schnek, A. G., and Léonis, J. (1975). Probable Identity of the Beta Chains from the Two Chicken Hemoglobin Components. *Biochimie*, 57, 843-844.

Villeponteau, B. and Martinson, H. (1981). Isolation and Characterization of the Complete Chicken Beta-Globin Gene Region: Frequent Deletion of the Adult Beta-Globin Genes in Lambda. *Nucl. Acids Res.*, 9, 3731-3746.

Williams, J. G., Kay, R. M., and Patient, R. K. (1980). The Nucleotide Sequence of the Major Beta-Globin mRNA from *Xenopus laevis*. *Nucl. Acids. Res.*, 8, 4247-4258.

REGULATION OF GENE EXPRESSION OF A DNA BACTERIOPHAGE AT THE TRANSLATIONAL LEVEL

Jim Karam

Department of Biochemistry
Medical University of South Carolina
171 Ashley Avenue
Charleston, South Carolina 29425

I. INTRODUCTION

The infection of *Escherichia coli* with bacteriophage T4 results in the cessation of all host macromolecular biosynthetic processes: DNA, RNA, and protein. The genome of T4 is a linear DNA duplex that encodes more than a hundred protein species, many of which are essential for productive viral replication.

Transcription of this genome begins immediately after its entry into the host cell. Initially, the host RNA polymerase recognizes a number of transcription initiation sites (promoters) on the T4 DNA and transcribes sets of genes that encode a variety of enzymes and other proteins that support viral DNA replication and that activate the transcription of additional sets of essential phage genes. Some of these proteins bind to the host RNA polymerase and endow it with new specificity for promoters that cannot be utilized *in vivo* by the unmodified host transcriptional enzymes. The outlines of the temporal sequence of T4 phage gene expression are known, but many intriguing details and questions remain unresolved. The reader is referred to the recent review by Rabussay and Guiduschek (1977) for a comprehensive account on the role of transcriptional mechanisms in T4 development.

The vast majority of T4 transcripts exhibit the high instability characteristic of prokaryotic mRNA (Astrachan and Volkin, 1958; Brenner, Jacob and Meselson, 1961). This has suggested that translational control mechanisms do not play a major role in determining the levels of phage proteins in the infected cells. Indeed, the conventional view of gene regulation in DNA prokaryotes follows the outlines of the Jacob and Monod (1961) model for control of the *lactose* operon in *E. coli*. In this model, the mRNA has a high-turnover rate and levels of *lac* enzymes (e.g. β-galactosidase) are determined by the amounts of short-lived mRNA made available by regulated transcription. In recent years, however, it has become evident that some prokaryotic mRNA species are intrinsically very stable and that the translation of some of these transcripts can be modulated in response to physiological conditions. A relatively well-studied example is the mRNA of T4 gene 32, the structural gene for the phage helix-destabilizing protein (Alberts and Frey, 1971). This protein binds preferentially, and cooperatively, to single-stranded nucleic acid. It was recently found that the gene 32 protein can regulate its own synthesis *(autogenous control)* by interacting, in a specific fashion, with its mRNA (Lemaire, Gold and Yarus, 1978). Translation of gene 32 mRNA is *repressed* in the presence of excess gene 32 protein and is *derepressed* under conditions of depleted supply of the protein (Gold, O'Farrell, and Russel, 1976; Krisch, Bolle, and Epstein, 1974; Russell et al., 1976). Other examples of prokaryotic mRNAs that are specifically regulated include the genomic RNA of RNA phages (Lodish, 1975) and the polycistronic transcripts of some of the *E. coli* ribosomal genes (Yates et al., 1980; Dean and Nomura, 1980; Brot et al., 1980). It is not clear if these documented cases of translational control signify types of mechanisms that are available for control of the various classes of cellular mRNA or if they merely represent exceptions that have survived evolutionary paths that favored transcriptional control for the prokaryotes. This report will deal with existence of a prokaryotic translational control process that is intimately associated with mRNA turnover and that is directed at several essential (and sometimes functionally unrelated) transcripts of the T4 virus. Such control processes may play a key role in the regulation of DNA viruses.

Mutations in the regA gene of phage T4 lead to overproduction of a small number of the proteins that are synthesized during the early stages of phage development (Karam and Bowles, 1974; Karam, McCulley and Leach, 1977; Sauerbier and Hercules, 1973; Wiberg et al., 1973). These mutations do not affect RNA synthesis; rather, the protein overproduction is accompanied by a prolonged functional lifetime for the corresponding mRNA. Cardillo et al. (1979) showed that the regA gene codes for a small protein (10-13,000 daltons). This protein seems to function as a translational inhibitor for several phage-derived transcripts, but no direct evidence is available on its mode of action. Possibly, the regA protein initiates mRNA breakdown directly, e.g. it may be a ribonuclease. Alternatively, it may inhibit ribosome loading on mRNA and mRNA breakdown may follow as a consequence of this inhibition. The evidence to be presented here is consistent with both models, but favors the notion that the regA protein recognizes a target on mRNA and *represses* translation by inhibiting initiation by ribosomes. The effects of T4 *regA* genetic lesions will be reviewed with special emphasis on localization of the mRNA target for recognition by regA protein.

II. THE T4 regA GENETIC REGION

Mutations in the T4 regA gene were discovered in two types of phage mutant searches. Wiberg et al. (1973) isolated the T4 *regA SP62* lesion as a phage mutant defective in its ability to degrade the host DNA. Such mutants are abnormally sensitive to hydroxyurea (Hercules et al., 1971). The T4 *regA R9* mutation was isolated by Karam and Bowles (1974) as a phage lesion that enhanced the growth of leaky mutants of specific T4 genes, e.g. 62 (a DNA replication function). The T4 *R9* mutant is also hydroxyurea-sensitive (HU^s), and it has been possible to map both *SP62* and *R9* relative to other T4 genetic lesions in genetic crosses between these HU^s mutants and nonsense mutants of various essential genes of the phage. These genetic mapping experiments placed the *regA* locus in the midst of a genetic cluster that is known to control replication of the phage DNA (Karam and Bowles, 1974; Cardillo et al., 1979). T4 *regA* mutants, however, do not exhibit any defects in replication although, as will be shown below, they appear to regulate the intracellular levels of some of the phage-induced DNA replication proteins.

Fig. 1 is a schematic diagram of the T4 chromosomal segment in which the regA gene maps. Information for this map was derived from the genetic crosses mentioned above and from restriction enzyme analyses of cloned fragments of T4 DNA. The restriction map shown in Fig. 1 is based largely on results from our laboratory and was matched to the genetic map by the use of "marker rescue" tests between normal T4 genetic sequences cloned bacterial plasmids and allelic mutant sequences present on infecting phage strains.

III. THE EFFECTS OF T4 regA MUTATIONS ON PHAGE-INDUCED PROTEIN SYNTHESIS

In normal infections of *E. coli* with T4 the rates of synthesis of "early" phage mRNAs decline substantially within a few minutes after the onset of phage DNA replication and of the initiation of transcription of the "late" phage gene functions (reviewed by Rabussay and Geiduschek, 1977). The rates of synthesis of T4 "early" proteins show a concomitant decline, reflecting the high instability of the phage-induced transcripts. In infections with T4 *regA* mutants, on the other hand, the phage continues to synthesize some of the "early" proteins despite the shutoff in "early" transcription (Karam and Bowles, 1974). The results of experiments that demonstrate the effects of T4 *regA* lesions are depicted in Fig. 2. In Fig. 2A it is shown that at late times after infection with the T4 *regA* mutant *R9*, some of the phage "early" proteins (e.g. prIIA, prIIB, p42, 45, p63, etc.) are synthesized at high rates relative to the rates seen with *regA*$^+$ phage. The synthesis of most T4 "early" proteins is unaffected by *regA* lesions. The example presented in Fig. 2B demonstrates that the *regA*-mediated overproduction of those few "early" phage proteins is due to a *posttranscriptional* rather than a *transcriptional* effect. In this experiment (Fig. 2B) phage-infected cells were allowed to synthesize T4 "early" mRNA for a few minutes and then they were treated with an RNA polymerase inhibitor. Under such conditions, it is possible to monitor the rates of decline in protein synthesizing capacity (functional decay) of individual phage-induced transcripts (Karam and Bowles, 1974). T4 *regA* mutations result in an increased mRNA functional lifetime for those proteins that are overproduced as a consequence of these lesions. We have also demonstrated that the increased mRNA functional stabilization in *regA*$^-$ infections is accompanied by decreased RNA degradation (Karam et al., 1977; Gerald and Karam, in preparation). So, the T4 regA gene function appears to be involved in the regulation of turnover and utilization of a specific subclass of phage-induced mRNA species.

IV. MOLECULAR NATURE AND FUNCTION OF THE T4 regA GENE PRODUCT

Genetic studies by Karam et al. (1977) suggested that the T4 regA gene product is a diffusable protein. Recently, Cardillo et

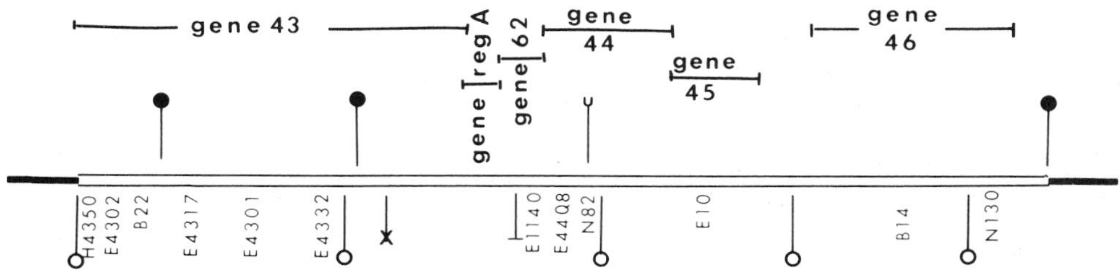

Fig. 1. A schematic diagram showing the location of T4 gene regA in relation to its neighboring genes on the phage chromosome. The map positions of the specific T4 mutations shown (H4350, E4302, B22, E4317, E4301, E4332, E1140, E4408, N82, E10, B14, and N130) was determined by marker rescue experiments in which phage carrying a mutation under study was used to infect bacteria containing a plasmid into which a segment of T4 DNA was cloned. The production of wild-type recombinants in the infection was indicative of the presence on the plasmid of the T4 wild-type DNA sequence allelic to the phage mutation being "rescued". The vertical lines that bear different symbols at their ends designate the approximate locations of cleavage sites by different restriction enzymes: ●, EcoR1; O, Hind III; X, XhoI, T, Pst I; and U, Hinc II. The restriction sites were determined by analysis of enzyme cleavage products of a variety of T4 DNA clones by agarose gel electrophoresis and by using Hae II, Hae III and Hinc II digests of ØX 174 DNA as sizing standards. Details of the cloning of T4 DNA segments in plasmids and the analysis of size and genetic content of this DNA will be reported elsewhere.

al. (1979) identified this protein by SDS-gel electrophoretic analyses of radioactive extracts of T4-infected cells. We have also identified this protein in studies on the molecular cloning of T4 genes and their expression under control of promoters in cloning vehicles. Dr. N. E. Murray at the University of Edinburgh, Scotland, cloned (in a lambdoid phage vehicle), a large portion of the T4 chromosomal segment diagrammed in Fig. 1 (personal

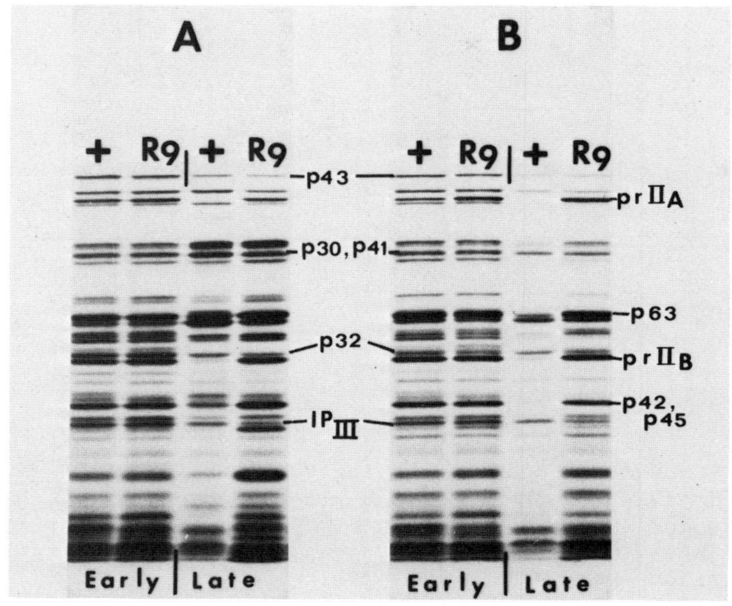

Time period after infection

Fig. 2. Reproductions of autoradiograms showing the effects of the T4 *regA* mutation *R9* on phage-derived "early" protein synthesis. In the experiment for panel A samples of infected cells were labelled with a ^{14}C-labelled mixture of L-amino acid at 6-11 min (Early time period) and 36-41 min (Late time period) after infection. In the experiment for panel B, the drug rifampicin (final concentration 200 µg/ml) was added to cultures at 5 min after infection and samples of the drug-treated cultures were labelled with the ^{14}C-amino acids at 8-13 min (Early time period) and 25-30 min (Late time period) after infection. The ^{14}C-labelled samples were subsequently used for SDS-gel electrophoresis and autoradiography according to methods detailed previously (Karam et al., 1977). The designation "R9" refers to infections with the T4 *regA R9* mutant; the designation "+" refers to infections with phage carrying the wild-type allele for *R9*. All phage strains used carried additional mutations that prevented phage DNA replication and the expression of "late" phage functions (Karam and Bowles, 1974). Some of the phage-induced proteins that are unaffected by *regA*$^-$ mutations are listed in the space between the two panels of the figure. Some of the proteins that are overproduced in *regA*$^-$ infections are listed on the right-hand side of panel B.

communication). We analyzed such a clone (named phage 761-4) made available to us by Dr. Murray and demonstrated expression of T4 genes 45, 44, 62, regA, and 43. An example of our results with this clone are shown in Fig. 3A. The identities of the T4 proteins made by clone 761-4 were established by comparing extracts of cells infected by phage 761-4 with a variety of extracts from cells infected with various T4 mutants that lacked the ability to synthesize specific protein species. In the case of the regA protein, we examined T4 *regA*$^+$ and several T4 *regA*$^-$ mutants. Some results are shown in Fig. 3B. These results confirm the observations made by Cardillo et al. (1979) that the regA protein is small (10-13,000 daltons) and that some defective *regA* peptides are overproduced along with the other proteins that respond to *regA*-mediated regulation. That is, the *regA* protein seems to regulate its own synthesis (autogenous regulation).

Two classes of models have been proposed to account for the selective partial inhibition of synthesis of certain T4 "early" proteins in *regA*$^+$ infections (Karam et al., 1977; Cardiollo et al., 1979). One type of model envisages the regA protein as a ribonuclease or other RNA-modifying enzyme that initiates RNA degradation. In this model, inhibition of translation ensues as a consequence of mRNA decay. The other type of model proposes that the regA protein interferes with the translation of certain mRNA species by interfering with the selectivity of ribosomes, i.e. with recognition of these mRNAs. In this model the regA protein acts as a "repressor" of translation and selective mRNA decay ensues as a consequence of a loss of protection of mRNA by ribosomes. Both types of models envisage the specificity of regA-mediated regulation to reside in specific nucleotide "target" sequences within the regulated mRNA's. The results to be described in the next section will deal with the identification of such a "target" sequence.

V. GENETIC LOCALIZATION OF A "regA TARGET" SEQUENCE ON THE T4 rIIB mRNA

The rIIB protein of T4 phage is known to be hyperproduced in *regA*$^-$ infections (e.g. Fig. 1). The structural gene for this protein has been well-characterized genetically (Barnett et al., 1967, Benzer, 1961; Nelson et al., 1981) as well as with regards to nucleotide sequence within the region that encodes elements of rIIB mRNA necessary for the initiation of translation (Nelson et al., 1981; Pribnow et al., 1981). In collaboration with L. Gold and B. S. Singer (University of Colorado), we screened a large number of partial deletions and other mutations in the T4 rIIB gene for loss of sensitivity of rIIB transcript to regulation by the phage regA function, i.e. for loss or alteration of the

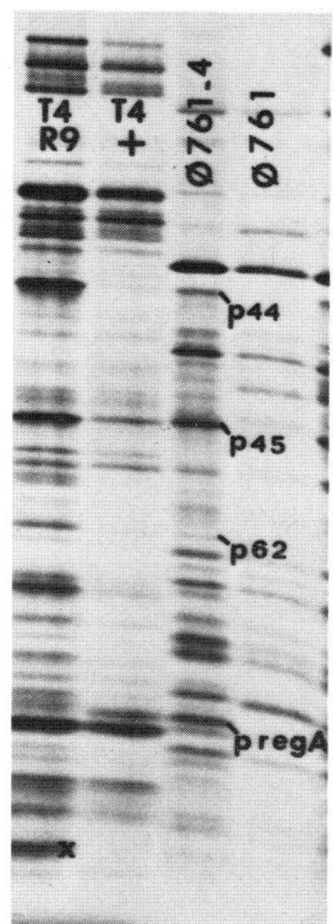

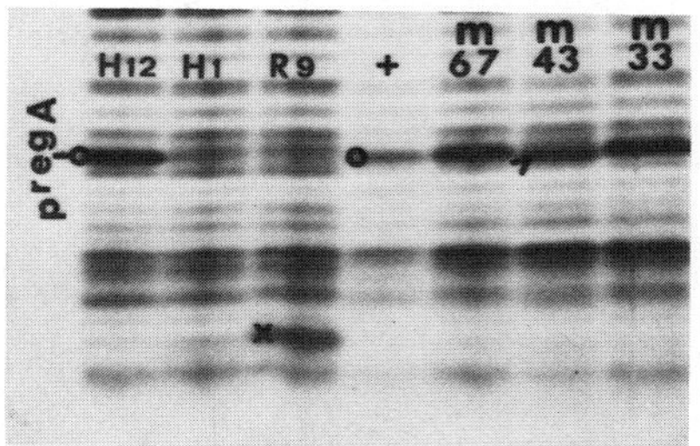

Fig. 3. Identification of the T4 *regA* protein by gel electrophoretic assays. The electrophoretic patterns of ^{35}S-labelled proteins extracted from cells infected with the lambdoid strains Ø 761 and Ø 761-4 were compared to the patterns of ^{14}C-labelled proteins from cell cultures that had been infected with a variety of T4 mutants. Strain Ø761 is the lambdoid cloning vector into which a segment of T4 DNA was cloned, resulting in Ø 761-4 (N. Murray, personal communication). In panel A, identification of the regA protein from T4 wild-type *(regA$^+$)* infections was accomplished via comparisons between T4 *regA R9*, T4 *regA$^+$*, Ø 761-4, and Ø 761. The experiment for panel B compared the *regA* proteins that are made by different T4 *regA* mutants. Note that some *regA* mutants overproduce their defective *regA* protein, indicating autogenous regulation (Cardillo et al., 1979). Experimental conditions were similar to those described in Fig. 2, except that in infections with the lambdoid phages UV-irradiated cells were used and inorganic $^{35}SO_4$ was the protein labelling isotope while ^{14}C-amino acids were used to label the T4-infected cells.

"*regA* target sequence". Our analyses (Karam et al., 1981) led to the identification of a 7-10 nucleotide sequence that overlaps the initiator codon (AUG), that is contained within the ribosome-binding sequence of rIIB mRNA (Belin et al., 1978), and that, when altered, renders rIIB mRNA translation insensitive to inhibition by $regA^+$ function. Mutants of this sequence exhibited overproduction of mutant rIIB proteins and increased functional lifetimes of the mutant mRNAs that encoded these proteins (Karam et al., 1981). In studies by E. Spicer and W. Konigsberg (personal communications), the Hin III DNA fragment that encompasses T4 gene 45 and part of gene 44 (Fig. 1) was sequenced. Since the products of both of these genes are sensitive to *regA*-mediated regulation (Karam et al., 1979), we searched the Spicer-Konigsberg sequence for "targets" similar to the one identified as rIIB mRNA. We found two such targets, one in the gene 45 ribosomal initiation region and the other in the gene 44 ribosome initiation region. The sequences of these proposed targets are compared in Fig. 4. The genetic analysis with T4 gene rIIB ruled out the existence of more than one "target" for regA-mediated regulation on rIIB mRNA (Karam et al., 1981). So, the T4 *regA* gene function appears to be related only to the process of initiation of translation.

5'.....GGAAAAUU(AUG)UACAAUAUUAAA.....3' gene rIIB

5'.....AAAUUAA(AUG)AAACUGUCU.....3' gene 45

5'.....AGGAAAUU(AUG)AUUACUGUAAAGAA.....3' gene 44

Fig. 4. A comparison of *regA* "target" sequences (underlined) in the translation initiation regions of T4 genes rIIB, 45, and 44. The rIIB nucleotide sequence for translation initiation was determined by Belin et al. (1979) and Pribnow et al. (1981). Initiator AUG codons are enclosed in parentheses. The nucleotide sequences for the genes 45 and 44 initiator regions were determined by E. Spicer and W. Konigsberg (personal communications). Genetic identification of the rIIB *regA* "targets" in gene 45 and 44 are inferred from nucleotide sequence data and have not yet been confirmed genetically.

VI. SUMMARY AND CONCLUSIONS

It is generally thought that translational control of gene expression exists primarily in eukaryotic organisms where the sites of transcription and protein synthesis are usually separated by physical barriers and where mRNA populations typically exhibit a high degree of metabolic stability (see Ochoa and deHaro, 1979). In contrast, the typical prokaryotic mRNA is metabolically unstable and prokaryotes are generally thought to use transcriptional control mechanisms, on the most part, to regulate the synthesis of their proteins (Jacob and Monod, 1961). Rapid mRNA turnover may provide an efficient means of recycling ribonucleotide precursors in organisms such as *E. coli* or in T4-infected *E. coli* where the needs for specific gene products vary rather frequently during the short growth cycle of the organism. On closer examination, however, a prokaryotic mRNA population, from virtually any source, can exhibit a wide range of intrinsic stabilities among its individual species. In T4-infected *E. coli*, for example, 5-10% of the protein species that are synthesized during the early phases of phage growth are encoded by metabolically stable transcripts (Krisch et al., 1974; Karam et al., 1977), and transcripts made during late times after phage infection appear to be more stable than "early" transcripts (Karam et al., 1977). So, the differential susceptibilities of mRNA species to intracellular degradative enzymes may constitute one means by which the levels of individual cellular proteins are controlled. Differential mRNA decay in bacteria has been a subject of study for several years and some instructive explanations have emerged. It appears that degradation of a transcript occurs by an orderly process that begins with cleavage(s) near the 5'-phosphoryl terminated end of the mRNA and proceeds towards the 3'-hydroxyl terminated end (Morikawa and Imamoto, 1969; Morse et al., 1969). The association of ribosome with the mRNA leads to protection against degradative activities. This has been shown to be the case for *E. coli* (Morse, 1970) and T4 transcripts (Walker et al., 1976). Ribosomes are known to translate different mRNAs at different efficiencies and so, it is possible that the observed differential decay rates among mRNA species in a cell simply reflect differences in the ribosome binding regions of these mRNAs. Our work on the T4 *regA* gene function indicates that other factors besides ribosomes and intrinsic mRNA structure can be involved in controlling the rate of inactivation of a transcript.

The studies reported here and elsewhere (Karam and Bowles, 1974; Karam et al., 1977; Karam et al., 1981) demonstrate that a small protein, the T4 *regA* gene product, is involved in control of translation of a subpopulation of T4 mRNA species. Mutational alterations of this gene product result in selective *translational derepression* of synthesis of several phage-induced proteins, this derepression is accompanied by increased stabilization of the mRNAs. One of these mRNAs (the T4 rIIB mRNA) was shown to harbor a short nucleotide sequence (7-10 bases long) in its ribosome-entry domain that determines sensitivity of this mRNA to inhibition by *regA*$^+$ function. Because of the small size of the *regA* protein and the small size of sequence for *regA*$^+$-mediated inhibition of rIIB translation, we propose that the *regA* protein interacts directly with the mRNA within the translation initiation region. The *regA* "target" sequence that we have identified in T4 rIIB in RNA may also be present in the translation initiator regions of two other regA-regulated mRNAs: the transcripts for T4 genes 45 and 44 (Spicer and Konigsberg, personal communications). Is the *regA* protein a specific ribonuclease or does it affect mRNA decay indirectly by "binding" to the sequence and preventing ribosomes from initiating? The small size of the "target" sequence for *regA*-mediated regulation makes it seem likely that it will be found in other mRNA species and at other locations in rIIB mRNA. If the *regA* protein were a ribonuclease that specifically recognized this sequence, then unwanted RNA decay would occur. Furthermore, the location of the "target" sequence is very suggestive of an involvement of the regA protein in regulation of *initiation* of translation. For these reasons we now favor a model that views the *regA* protein as a repressor of translation that acts in a manner analogous to transcriptional repressors, i.e. by binding to its target sequence. We are currently purifying *regA* protein and its target sequences and we will be able to differentiate between the two types of activities of this protein by direct *in vitro* assay.

If the *regA* protein is indeed an RNA-binding regulator of translation, then it would be unique among such regulatory proteins. Our genetic studies with T4 rIIB gene (Karam et al., 1981) gave no indication that secondary or higher orders of RNA structure are relevant to regulation by the *regA* function; only *sequence* changes within the identified "target" exhibited derepression. Other translational regulatory proteins recognize *structural* domains within their target regions. For example, the coat protein of RNA phages sees a hairpin loop structure that overlaps the initiation codon for the replicase cistron (the repressed gene) and the T4 gene 32 protein may recognize a region on its mRNA that avoids formation of secondary structure (Borisova et al., 1979; Gralla et al., 1974; Steitz, 1974; Russel et al., 1976; Lemaire et al., 1978; Krisch et al., 1978). In addition, the *regA* protein appears to be a global regulator of translation in that some of the functions controlled by this protein are not related to one another and several are genetically unlinked. Translational regulatory mechanisms of this type have not been described before.

VII. REFERENCES

Alberts, B. M., and Frey, L. (1970). T4 bacteriophage gene 32: a structural protein in the replication and recombination of DNA. *Nature*, 227, 1313-1318.

Astrachan, L. and Volkin, E. (1958). Properties of ribonucleic acid turnover in T2-infected *Escherichia coli*. *Biochim. Biophys. Acta*, 29, 536-544.

Barnett, L., Brenner, S., Crick, F. H. C., Shulman, R. G., and Watts-Tobin, R. J. (1967). Phase-shift and other mutants in the first part of the rIIB cistron of bacteriophage T4. *Phil. Trans. Royal Soc. London, Series B*, 252, 487-560.

Belin, D., Hedgpeth, J., Selzer, G. B., and Epstein, R. H. (1979). Temperature-sensitive mutation in the initiation codon of the rIIB gene of bacteriophage T4. *Proc. Natl. Acad. Sci. USA*, 76, 700-704.

Benzer, S. (1961). On the topography of the genetic fine structure. *Proc. Natl. Acad. Sci. USA*, 47, 403-415.

Borisova, G. P., Volkova, T. M., Berzin, V., Rosenthal, G., and Gen, E. J. (1979). The regulatory region of MS2 phage RNA replicase cistron. IV. Functional activity of specific MS2 RNA fragments in formation of the 70S initiation complex of protein biosynthesis. *Nuc. Acid Res.*, 6, 1761-1774.

Brenner, S., Jacob, F., and Meselson, M. (1961). An unstable intermediate carrying information from genes to ribosomes for protein synthesis. *Nature*, 190, 576-581.

Brot, N., Caldwell, P., and Weissbach, H. (1980). Autogenous control of *Escherichia coli* ribosomal protein L10 synthesis *in vitro*. *Proc. Natl. Acad. Sci. USA*, 77, 2592-2595.

Cardillo, T. S., Landry, E. F., and Wiberg, J. S. (1979). regA protein of bacteriophage T4D: identification, schedule of synthesis, and autogenous regulation. *J. Virol.*, 32, 905-916.

Dean, D. and Nomura, M. (1980). Feedback regulation of ribosomal protein gene expression in *Escherichia coli*. *Proc. Natl. Acad. Sci. USA*, 77, 3590-3594.

Gold, L., O'Farrell, P. Z., and Russel, M. (1976). Regulation of Gene 32 expression during bacteriophage T4 infection of *Escherichia coli*. *J. Biol. Chem.*, 251, 7251-7262.

Gralla, J., Steitz, J. A., and Crothers, D. M. (1974). Direct physical evidence for secondary structure in an isolated fragment of R17 bacteriophage mRNA. *Nature*, 248, 204-208.

Hercules, K., Munro, J. L., Mendelsohn, S., and Wiberg, J. S. (1971). Mutants in a nonessential gene of bacteriophage T4 which are defective in the degradation of *Escherichia coli* deoxyribonucleic acid. *J. Virol.*, 7, 95-105.

Jacob, F. and Monod, J. (1961). Genetic regulatory mechanisms in the synthesis of proteins. *J. Mol. Biol.*, 3, 318-358.

Karam, J. D. and Bowles, M. G. (1974). Mutation to overproduction of bacteriophage T4 gene products. *J. Virology*, 13, 428-438.

Karam, J., McCulley, C., and Leach, M. (1977). Genetic control of mRNA decay in T4 phage-infected *Escherichia coli*. *Virology*, 76, 685-700.

Karam, J., Bowles, M., and Leach, M. (1979). Expression of bacteriophage T4 genes 45, 44, and 62. 1. Discoordinate synthesis of the T4 45- and 44- proteins. *Virology*, 94, 192-203.

Karam, J., Gold, L., Singer, B. S., and Dawson, M. (1981). Translational regulation: Identification of the site on bacteriophage T4 rIIB mRNA recognized by the regA gene function. *Proc. Natl. Acad. Sci. USA*, in press.

Krisch, H. M., Bolle, A., and Epstein, R. H. (1974). Regulation of synthesis of bacteriophage T4 gene 32 protein. *J. Mol. Biol.*, 88, 89-104.

Krisch, H. M., Houwe, G. V., Belin, D., Gibbs, W., Epstein, R. H. (1977). Regulation of the expression of bacteriophage T4 genes 32 and 43. *Virology*, 78, 87-98.

Lemaire, G., Gold, L., and Yarus, M. (1978). Autogenous translational repression of bacteriophage T4 gene expression *in vitro*. *J. Mol. Biol.*, 126, 73-90.

Lodish, H. F. (1975). Regulation of *in vitro* protein synthesis by bacteriophage RNA by tertiary structure. In "RNA Phages" (N. D. Zinder, ed.) pp. 301-318. *Cold Spring Harbor Laboratory, Cold Spring Harbor, N.Y.*

Morikawa, N. and Imamoto, F. (1969). On the degradation of messenger RNA for the tryptophan operon in *Escherichia coli*. *Nature (London)*, 223, 37-40.

Morse, D. E. (1970). "Delayed-early" mRNA for the tryptophan operon? An effect of chloramphenicol. *Cold Spring Harbor Symp. Quant. Biol.*, 35, 495-496.

Morse, D. E., Mosteller, R., Baker, R. F., and Yanofsky, C. (1969). Degradation of tryptophan messenger. *Nature (London)*, 223, 40-43.

Nelson, M. A., Singer, B. S., Gold, L., and Pribnow, D. (1981). Mutations that detoxify an aberrant T4 membrane protein. *J. Mol. Biol.*, 149, 377-403.

Ochoa, S. and deHaro, C. (1979). Regulation of protein synthesis in eukaryotes. *Ann. Rev. Biochem.*, 48, 549-580.

Pribnow, D., Sigurdson, D. C., Gold, L., Singer, B., Napoli, C., Brosius, J., Dull, T. J., and Noller, H. F. (1981). The rII cistrons of bacteriophage T4: DNA sequence around the intercistronic divide and positions of genetic landmarks. *J. Mol. Biol.*, 149, 337-376.

Rabussay, D. and Geiduschek, E. P. (1977). Regulation of gene action in the development of lytic bacteriophages. In "Comprehensive Virology" (Fraenkel-Conrat, H., and Wagner, R. R., eds.) Vol. 8, pp. 1-196, *Plenum Press, New York and London.*

Russel, M., Gold, L., Morrisset, H., and O'Farrell, P. Z. (1976). Translational, autogenous regulation of gene 32 expression during bacteriophage T4 infection. *J. Biol. Chem.*, 251, 7263-7270.

Sauerbier, W., and Hercules, K. (1973). Control of gene function in bacteriophage T4. IV. Post-transcriptional shut-off of expression of early genes. *J. Virol.*, 12, 538-547.

Steitz, J. A. (1974). Specific recognition of the isolated R17 replicase initiator region by R17 coat protein. *Nature*, 248, 223-225.

Walker, A. C., Walsh, M. L., Pennica, D., Cohen, P. S., and Ennis, H. L. (1976). Transcription-translation and translation-messenger RNA decay coupling: Separate mechanisms for different messengers. *Proc. Natl. Acad. Sci. USA*, 73, 1126-1130.

Wiberg, J. S., Mendelsohn, S., Warner, V., Hercules, K., Aldrich, C., and Munro, J. L. (1973). SP62, a viable mutant of bacteriophage T4D defective in regulation of phage enzyme synthesis. *J. Virol.*, 12, 775-792.

Yates, J. L., Arfsten, A. E., and Nomura, M. (1980). *In vitro* expression of *Escherichia coli* ribosomal protein genes: autogenous inhibition of translation. *Proc. Natl. Acad. Sci. USA*, 77, 1837-1841.

IMMUNOGLOBULIN GENETICS

An-Chuan Wang

Department of Basic and Clinical Immunology and Microbiology
Medical University of South Carolina
171 Ashley Avenue
Charleston, South Carolina 29425

I. INTRODUCTION

Vertebrates possess a highly developed defense mechanism, the immune system, which protects them from pathogenic foreign invaders (e.g., microorganisms). The immune system has two major compartments, humoral and cellular immunity. Immunoglobulins (Ig) are the central elements for both immune compartments; they are antibodies of the humoral compartment, and they also serve as antigen receptors on the membrane of both bone-marrow-derived (B) and thymus-derived (T) lymphocytes, although the "Ig" produced by T cells are different from the classical Ig produced by B cells (Marchalonis, 1980; Woodland, and Cantor, 1980).

The genetic control of Ig synthesis has long been intriguing. Antibodies are remarkably specific in their ability to distinguish antigenically foreign macromolecules from self constituents and to differentiate very subtle differences among intruding antigens. The differences between some related antigens are so slight that the number of structurally different antibodies a vertebrate can produce against an almost indefinite supply of antigens must be enormous; an educated guess of the lowest limit of this number is in the neighborhood of one million.

Antibodies can be differentiated serologically with monospecific antisera directed toward a variety of antigenic determinants of the Ig molecule. Four kinds of antigenic determinants are important in Ig genetics. These are isotypic, allotypic, idiotypic, and isoallotypic determinants. Isotypic determinants differentiate among the heavy (H) chain classes (e.g., γ, α, and μ) and subclasses (e.g., $\gamma 1$ and $\gamma 2$) as well as light (L) chain types (i.e., κ and λ) and subtypes (e.g., OZ and KORN λ chains). They are present in all normal individuals. Allotypic determinants are antigenic determinants specified by allelic genes. Each allotypic determinant may not be present in all normal individuals. Whether a given individual has or does not have a given allotype depends on the genetic constitution (i.e., genotype) of that individual. Allotypes generally reflect regular small differences, between individuals of the same species, in the amino acid sequences of otherwise similar Ig polypeptide chains. Idiotypic determinants are antigenic determinants unique to a specific population of Ig molecules, which are presumably synthesized by a single clone of Ig-producing cells. Idiotypic determinants are located inside and/or adjacent to the antigen-binding site; thus, they are related mainly to the hypervariable regions. An isoallotypic determinant is an antigenic determinant that behaves as an allotype within a specific subclass of H chain but also behaves as an isotype in at least one of the other H-chain subclasses. Therefore, isoallotypes are present in all normal individuals, but each of them acts as a genetic allele of a single C_H locus. Isoallotypic determinants are useful for tracing the phylogenetic relationship of Ig genes.

Immunoglobulin genetics has several unique features. First, there is the phenomenon of specific gene activation. Although a normal individual possesses hundreds or more genes for heavy-chain variable regions (V_H) and light-chain variable regions (V_L) as well as several genes for heavy-chain constant regions (C_H) and light-chain constant regions (C_L), each Ig-producing cell generally expresses only one gene in each of these four gene families (Steinberg et al., 1970). Secondly, in individuals heterozygous for a given Ig allotype, only one of the two alternative alleles is expressed on each mature B cell. This phenomenon is known as allelic exclusion (Hood et al., 1975; Wang and Wang, 1979). Thirdly, Ig genes defy the well-established rule of "one gene for one polypeptide chain", in that the variable (V) and constant (C) regions of each Ig polypeptide chain are encoded by separate structural genes (Hood, 1972; Wang, 1978). Finally, the expression of Ig structural genes appears to be under the control of as yet unidentified regulatory genes. Many of the so-called complex (or latent) allotypes (e.g., the rabbit V_H allotypes of the a group) may represent multiple closely link isotypes, but their expression is controlled by different alleles of a hypothetical regulatory gene (Wang, 1974; Strosberg, 1977). Since the genetics of Ig is a very wide subject, and to cover the entire area in a short review article would be extremely difficult, I will discuss only some of the most interesting recent developments below.

II. THE MOLECULAR BASIS FOR SPECIFIC GENE ACTIVATION AND ALLELIC EXCLUSION

The phenomena were noticed in the early 1960s, but it was not until the past 5 years that immunogeneticists began to understand the molecular mechanism governing their manifestation. Structural and genetic analyses demonstrated that the basic unit of each Ig molecule is encoded by at least four genes: V_H, C_H, V_L, and C_L. (Nucleotides encoding the J segment of both H and L chains and the D segment of the H chain are also considered as mini-genes by some immunologists.) A normal individual has large numbers (hundreds or more) of V_H and V_L genes and moderate numbers of C_H and C_L genes. These genes belong to three linkage groups, as illustrated in Fig. 1. In humans, the H-chain are located on chromosome 14, whereas K-and λ- chains genes are on chromosomes 2 and 22, respectively. In mice, the H-, K-, and λ-chain genes are on chromosomes k2, 6, and 16, respectively.

Nucleic acid analyses have shown that, within each linkage group, the V genes are far apart from the C genes on the chromosome in germ and embryonic cells. However, in differentiated lymphocytes of adults, one of the V genes is translocated on the neighborhood of a C gene (Hozymi and Tonegowa, 1976; Davis et al., 1980; Seidman et al., 1979). This V-C translocation (or more precisely V-J translocation, as shown in Fig. 2) is apparently responsible for the specific gene activation, because the cell becomes committed to the production of the particular Ig polypeptide chain encoded by the translocated V-C gene pair (or V-J-C gene segments). Current evidence suggests that this translocation generally occurs on only one of the two homologous chromosomes in a given Ig-producing cell (Joho and Weissman, 1980; Vaoita and Hojo, 1980). Thus, the allelic exclusion phenomenon can also be explained by the V-C (or V-J) joining event.

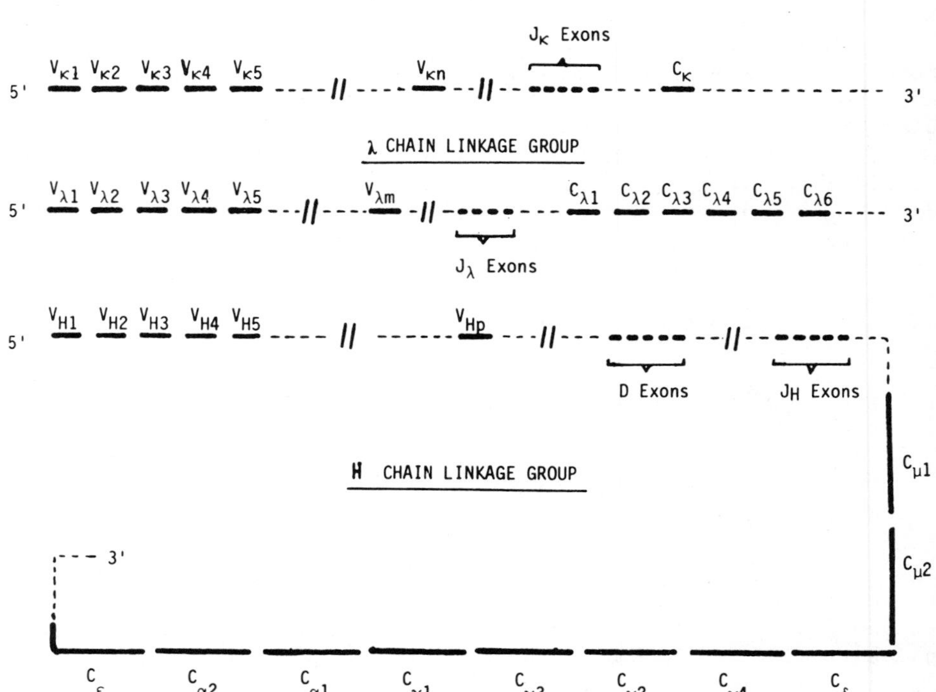

Fig. 1. Schematic representation of Ig structural genes in human. Each J exon encodes approximately 13 amino acid residues at the carboxyl-terminal end of the V region. Each D exon encodes a short stretch of amino acids equivalent to the third hypervariable region of the H chain.

Further nucleic acid studies have shown that Ig genes are split genes, in that deoxynucleotide sequences which are translated into amino acids are separated by untranslated intervening sequences. The former are called exons and the later introns. In the DNA, the borders of the introns are usually marked by GT and AG sequences. The genetic events leading to the synthesis of an Ig L chain is outlined in Fig. 2. In germ and embryonic cells the nucleotides encoding a full L chain are located at two large DNA fragments, which are far apart. One of the fragments contains the Leader (L) and V exons, which are separated by an intron. The second DNA fragment contains the J and C exons, which are separated by another intron. During differentiation of Ig-producing cells, a somatic translocation joins the

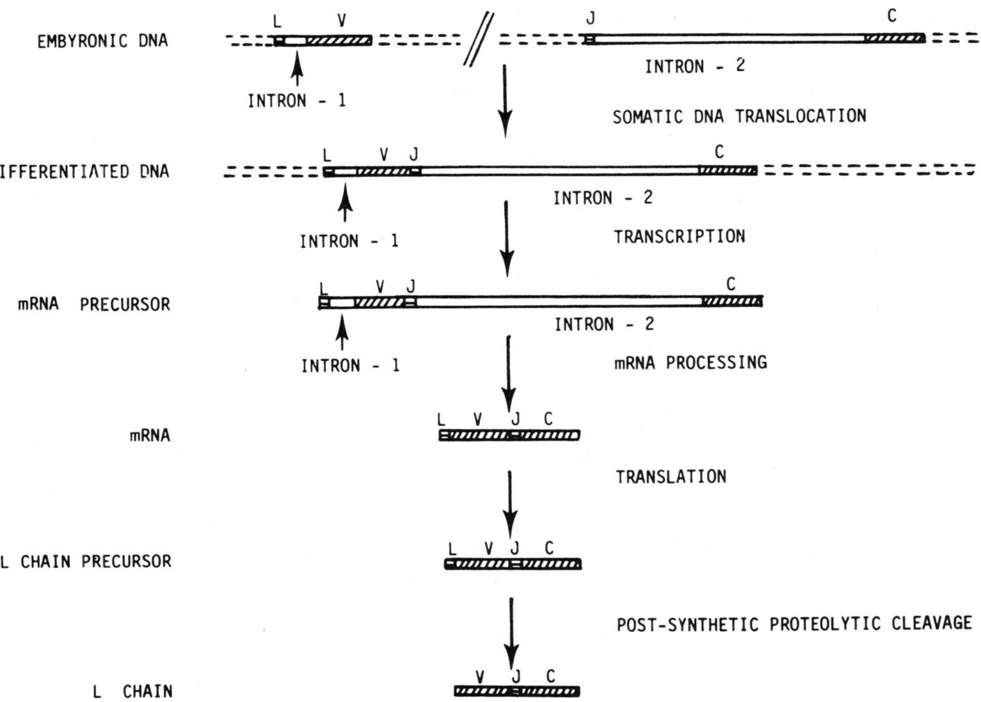

Fig. 2. Genetic events involved in the production of an Ig L chain. Shaded areas represent exons, white bars represent introns. The L exon codes for the leader piece.

V to the J. This translocation commits the cell and its progeny to the synthesis of the L chain specified by the particular V, J, and C exons involved. In differentiated cells, all deoxynucleotides from the leader exon to the C exon are transcribed into a precursor messenger RNA (mRNA), which is processed inside the nucleus to delete the introns before becoming a mature mRNA. The mature mRNA then goes to the polyribosome to direct the translation into an L chain precursor. A postsynthetic proteolytic cleavage cuts off the leader piece from the NH_2-terminal end of the precursor L chain, and turns the precursor L chain into a mature L chain.

The mechanisms that govern the H-chain synthesis are even more complicated. Based on amino acid sequence and serological data (Fig. 3a) on a monoclonal IgM and a monoclonal IgG from one patient, we have proposed a "genetic switch" hypothesis (Wang et al., 1970). This hypothesis suggest that during B-cell maturation, certain IgM-producing cells undergo a switch to the production of Ig belonging to another class (Fig. 3b). During the switch a given V_H gene, which was associated with a C μ gene for the production of IgM, is translocated to another C_H gene for the synthesis of the new Ig class, which retains the same idiotype and antigen-binding specificity as the parental IgM. This hypothesis was subsequently verified by nucleic acid analysis. In 1980, at least seven laboratories independently reported that two types of DNA rearrangments are involved in the synthesis of H chains other than the μ chain (Davis et al., 1980; Rabbitts et al., 1980; Ravetch et al., 1980; Cory et al., 1980; Coleclough et al., 1980; Sakano et al., 1980; Kataoka et al., 1980). The first rearrangement is equivalent to V-J translocation of the L chain, whereby a V_H exon is joined to a J_H exon adjacent to the C μ gene. The second rearrangement, the C_H switch, replaces the C μ gene with another C_H gene. A schematic outline of genetic events involved in H-chain production is shown in Fig. 4. It is generally believed that the initial V-J translocation leads to IgM production, whereas the C_H switch leads to the production of IgG, IgA, or IgE. The possible mechanism leading to IgD synthesis are not yet fully understood; they may involve either a C_H switch or mRNA processing or both (Wable et al., 1980; Lin et al., 1980; Moore et al., 1981).

Although the mechanism of the V-J joining and the C_H switch have not been determined, a majority of the experimental evidence is consistent with a "loop-out deletion" model (Honjo and Kataoka, 1978; Sakano et al., 1979; Rabbitts et al., 1980; Cory et al., 1980). Fig. 5 illustrates such a model using a mouse κ chain system as an example. DNA sequence studies revealed certain unusual stretches (approximately 6 to 7 nucleotides long) of DNA with palindrome (inverted repeat) sequences in the introns after the V exon and before the J exon. The complementary relation between the palindrome sequences (there are generally many palindromic pairs separated by spacers in each system, but only one pair is shown in Fig. 5 to illustrate the complementarity) allow them to become aligned to bring together widely separated V and J exons. It is conceivable that the paired palindromes may also serve as a recognition site for an enzyme (e.g., recombinase) that catalyzes the cleavage and the splicing steps, and that during these steps the V-J segments are fused and the nucleotides in between are deleted. The evidence in general favors the idea that such a mechanism is operative in B cells that are differentiating into plasma cells. There is also evidence against a simple deletion model (Nottenberg and Weissman, 1981). The observation by Couderc et al. (1979) that individual mouse B cells may occasionally produce two different antibody molecules argues against the

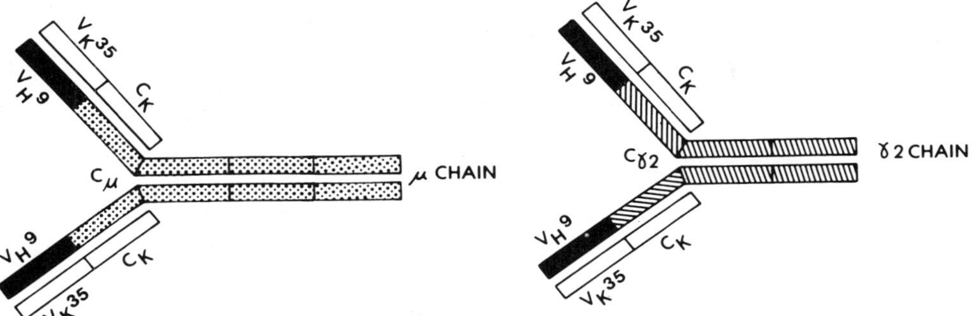

Fig. 3. (a) Schematic representation of the structural data for a monoclonal IgM(κ) and a monoclonal IgG2(κ) isolated from a single patient (Til). The two molecules have identical L chains and identical V_H regions (Wang et al., 1977), but their C_H regions have less than 30% amino acid sequence homology. Rhodamine-labeled anti-μ stained IgM-producing cells only, and fluorescein-labeled anti-γ stained IgG-producing cells only (Wang et al., 1969), but fluorescent anti-idiotype antibodies reacted with both populations of cells. (b) Diagramatic representation of the genetic switch hypothesis for B-cell differentiation. A given V_H gene (e.g., $V_H{}^9$) was initially translocated to the Cμ gene for the production of IgM. It was subsequently switched to the neighborhood of another C_H gene (e.g., Cγ2) for the production of another Ig class (From <u>Basic Immunogenetics</u>, 2nd edition, by H. H. Fudenberg, J. R. L. Pick, A. C. Wang, and S. D. Douglas, Oxford University Press, New York, 1978).

idea that the loop-out deletion occurs in all B cells, including those at an earlier development stage. We should not totally dismiss the following alternatives to a deletion mechanism: (a) the V-J joining or C_H switching may occur in such a manner that the intervening DNA is not always lost, and (b) crossing-over may occasionally occur between sister chromatids or homologous chromosomes. Moreover, the recognition sequences for different C_H genes differ from one another, and Davis et al. (1980) have suggested that C_H switching may be mediated by class-specific recognition sequences and class-specific switching enzymes.

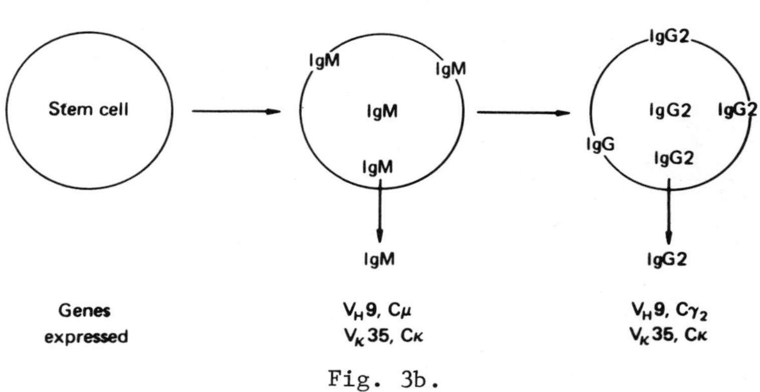

Fig. 3b.

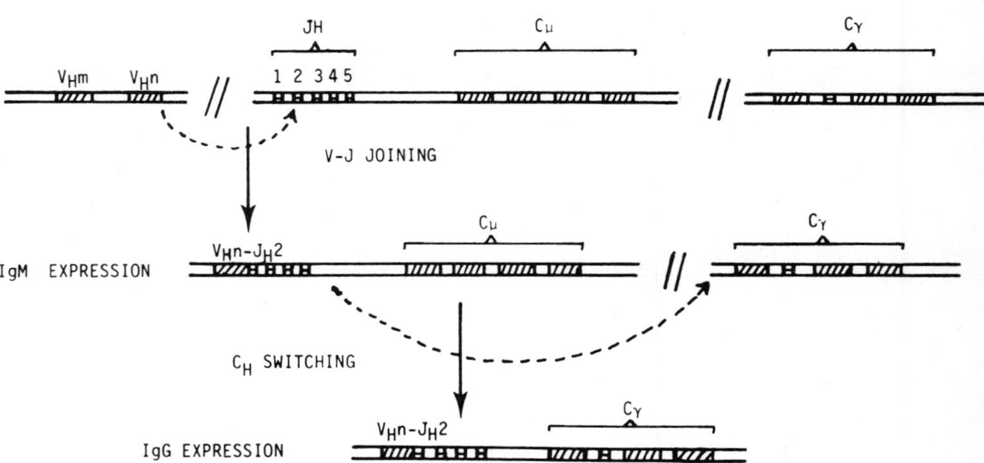

Fig. 4. Two types of DNA rearrangments during the differentiation of B-lymphocytes: (1) V-J joining (expression of the Cμ gene and (2) C_H switching (expression of other C_H genes. Simplified model for the expression of H-chain genes. At least two DNA rearrangements (i.e., V-J translation and C_H switching) must be involved. It is likely that a V-D or D-J joining event precedes the V-J joining.

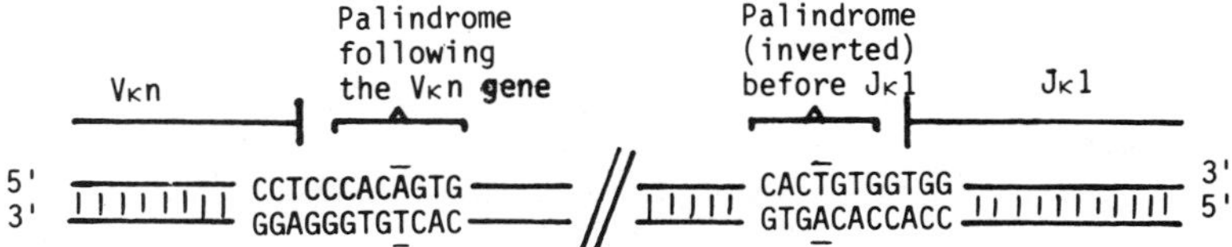

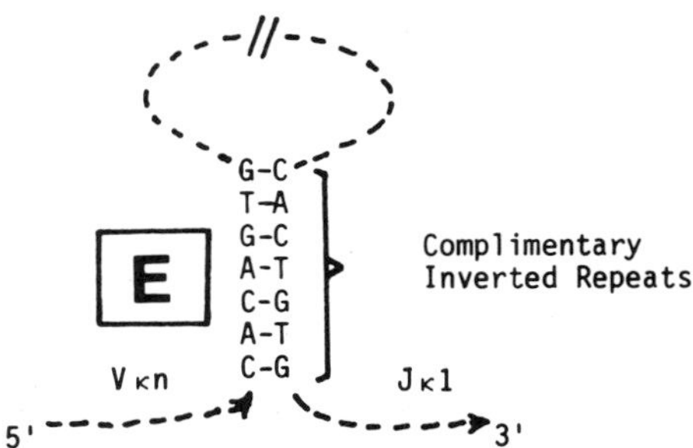

Fig. 5. A simplified model for V-J joining of L chain genes. The complimentary inverted repeats (palindormes) bring together widely separated V and J segments. An hypothetical enzyme (e.g. recombinase) then catalyzes the splicing event.

III. HOMOLOGY REGIONS (DOMAINS) AS UNITS FOR IMMUNOGLOBULIN EVOLUTION

Amino acid sequence analysis has shown that Ig polypeptide chains are composed of homology regions. Each homology region is approximatley 110 and 120 amino acid residues long and has an intrachain disulfide bond bridging two cysteine residues approximately 60 residues apart. Under physiological conditions, each homology region is folded into a compact globular structure called a domain (Gally and Edelman, 1972). It is generally believed that all structural genes coding for Ig polypeptide chains have evolved from a common ancester gene that coded for a polypeptide chain equivalent to one domain (Singer and Doolittle, 1966; Hill et al., 1966). Nucleotide sequence analyses indicate that each homology region is encoded by an exon (Sakano et al., 1979; Early et al., 1979). Exons encoding various C_H domains and the hinge region are separated by introns in the C_H gene. Thus, such C_H exons may serve as convenient units for unequal crossing-over between C-region genes.

This concept is supported by the pattern of deletion mutants in heavy-chain disease (HCD) proteins (Frangione and Franklin, 1975) and by the occurrence of hybrid H chains that contain C_H domains of more than one C-region subclass (Kunkel et al., 1969; Natvig and Kunkel, 1974; Tsuzukida et al., 1979). Recently, studies in our laboratory have also provided experimental evidence in favor of this concept. We have performed biochemical analyses of two human HCD proteins. One of them (Mia) was a γ3-chain deletion mutant with intact C_H2 and C_H3 domains but with both the V_H and C_H1 domains missing (Wang et al., 1978). The other (Cha) was a γ3-γ1 hybrid; it resembled the γ3 chain at the hinge region but resembled the γ1 chain at the C_H2 and C_H3 domains (Arnaud et al., 1981). Again, the V_H and C_H1 domains were deleted (Fig. 6).

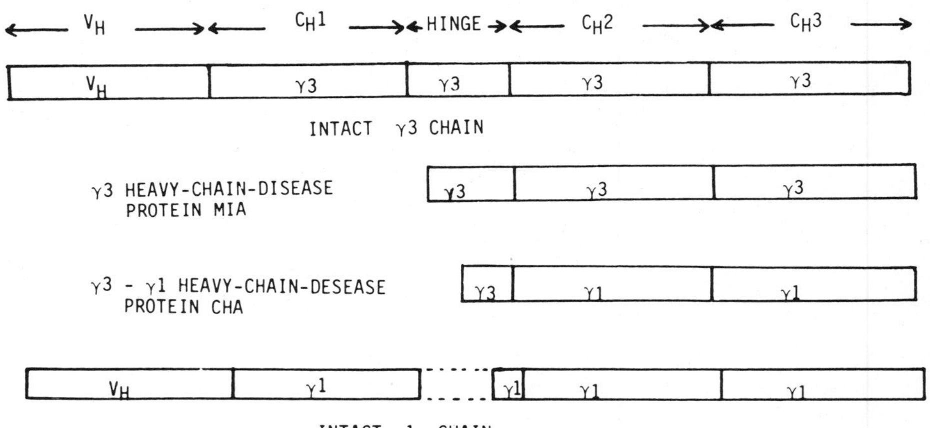

Fig. 6. Schematic representation of two heavy-chain disease proteins studied in our laboratory, both involving deletion of the entire V_H and C_H1 domains.

More interesting, we have found a hybrid Ig molecule (designated IgM/A) that reacts with both anti-IgM and anti-IgA. The hybrid molecule is not an IgM-IgA immune complex because it is smaller than pentameric (19S) IgM as judged by its elution profile from a Sephacryl S-300 column. Furthermore, SDS polyacrylamide gel electrophoresis showed that the hybrid IgM/A has only one homogeneous H chain, with a molecular weight of approximately 70,000

(Tung et al., 1981). The occurrence of a μ-α hybrid H chain provides indirect evidence that the Cμ gene may be closer to the Cα than to the Cγ genes on chromosome 14 in humans; this differs from the arrangement of C_H genes in mice.

The idea that Ig homology regions may have played an important role in evolution is also supported by comparison of C-region amino acid sequences of various mammalian γ chains. To date, complete amino acid sequences have been reported for three human γ1 chains (Edelman et al., 1969; Ponstingl and Hilschmann, 1972; Cunningham et al., 1971), one human γ2 chain (Wang et al., 1980), one rabbit γ chain (Fructer et al., 1970; Columb and Porter, 1975), one mouse γ1 chain (Milstein et al., 1974), one mouse γ2a chain (Fougereau et al., 1976), and one guinea pig γ2 chain (Brunhouse and Cebra, 1979). Extensive sequences have also been reported for another human γ2 chain (Connell et al., 1979) and pooled guinea pig γ1 chain (Brunhouse and Cebra, 1979). A comparison of these sequences is shown in Fig. 7. Substantial sequence homology is present among all of them; they are identical at over 35% of the C_H positons, excluding the hinge region (residue positions with identical amino acids in all nine γ chains compared are enclosed in boxes).

Fig. 8 shows a quantitative analysis of the present amino acid sequence homology among γ chains. The γ chains of the four mammalian species compared have approximately 50 to 75% sequence homology within their respective C_H domains when the hinge region is excluded. The human γ1 chain has similar degrees of sequence homology with mouse and guinea pig γ1 chains as with their γ2 chains. The same is also true when human γ2 chains are compared with various mouse and guinea pig γ-chain subclasses. Therefore, the use of similar nomenclature for γ-chain subclasses of different mammalian species should not be interpreted as being indicative of phylogenetic similarities, since the γ1 chains of each species show no closer evolutionary relationship with the γ1 chains of other species than with their γ2 chains. In terms of intraspecies comparison, the human γ2 chain has approximately 90% or more sequence homology with the human γ1 chain at all three corresponding C_H domains. In contrast, the mouse γ1 chain has only 61 to 73% sequence homology with the mouse γ2a chain at the three corresponding domains. The guinea pig γ1 chain has 91% sequence homology with its γ2 chain at the C_H1 domain (parallel to that between human γ-chain subclasses) but 70% and 64% homology, respectively, at the corresponding C_H2 and C_H3 domains (parallel to that between mouse γ-chain subclasses). These comparisons indicate that in man as well as in mouse, exons that encode all three C-region domains of a given γ chain have evolved in parallel in one germ-line DNA segment as one gene unit, whereas the exons coding for the C_H1 domains of guinea pig γ chains have evolved independently from those that encode the C_H2 and C_H3 domains of its γ chain. In guinea pigs, an additional rearrangement

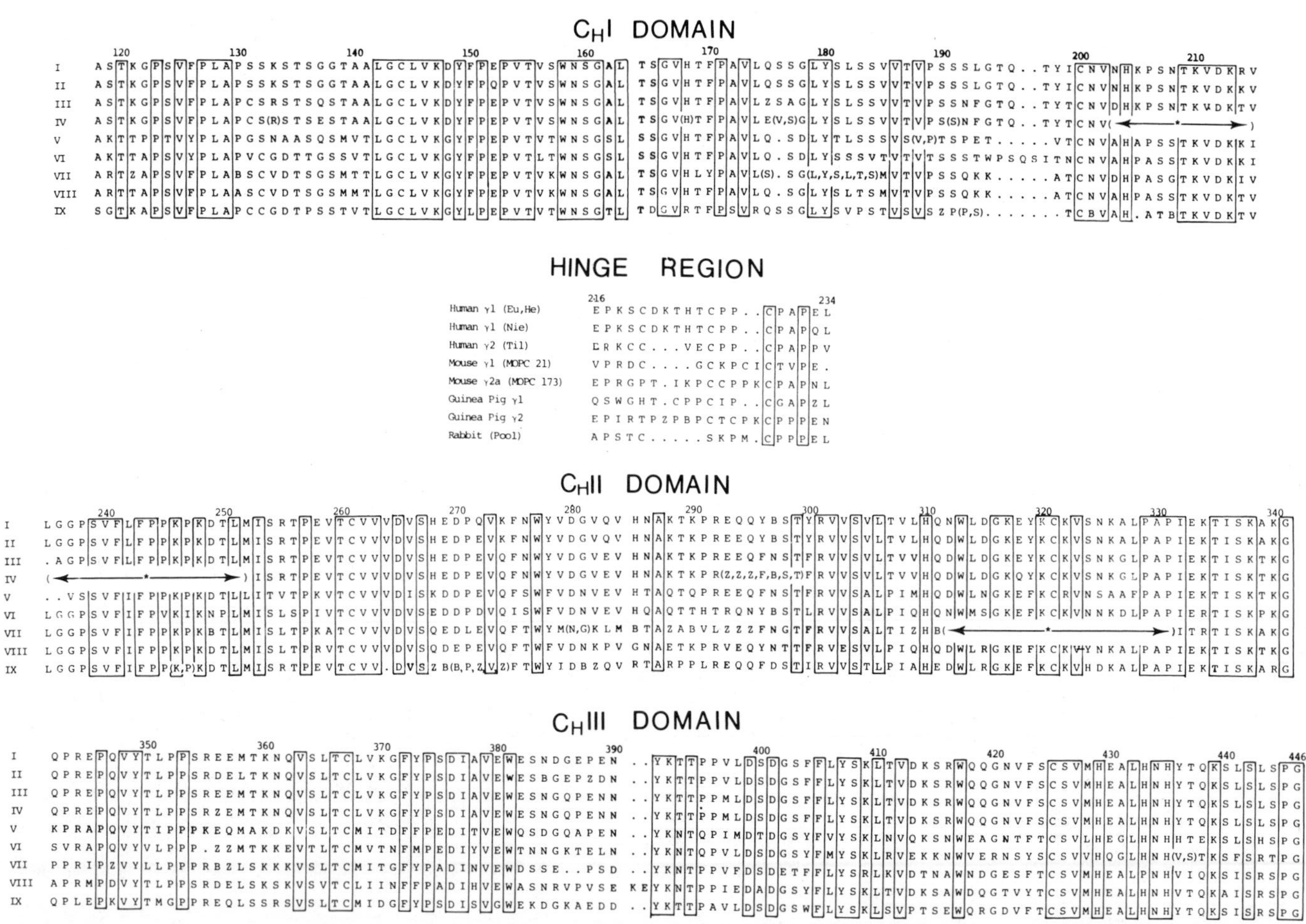
CHI DOMAIN
I ASTKGPSVFPLAPSSKSTSGGTAALGCLVKDYFPEPVTVSWNSGAL TSGVHTFPAVLQSSGLYSLSSVVTVPSSSLGTQ..TYICNVNHKPSNTKVDKRV
II ASTKGPSVFPLAPSSKSTSGGTAALGCLVKDYFPQPVTVSWNSGAL TSGVHTFPAVLQSSGLYSLSSVVTVPSSSLGTQ..TYICNVNHKPSNTKVDKKV
III ASTKGPSVFPLAPCSRSTSQSTAALGCLVKDYFPEPVTVSWNSGAL TSGVHTFPAVLZSAGLYSLSSVVTVPSSNFGTQ..TYTCNVDHKPSNTKVDKTV
IV ASTKGPSVFPLAPCS(R)STSESTAALGCLVKDYFPEPVTVSWNSGAL TSGV(H)TFPAVLE(V,S)GLYSLSSVVTVPS(S)NFGTQ..TYTCNV(←*→)
V AKTTPPTVYPLAPGSNAASQSMVTLGCLVKGYFPEPVTVTWNSGSL SSGVHTFPAVLQ.SDLYTLSSSVS(V,P)TSPET.....VTCNVAHAPSSTKVDKKI
VI AKTTAPSVYPLAPVCGDTTGSSVTLGCLVKGYFPEPVTLTWNSGSL SSGVHTFPAVLQ.SDLYSSSVTVTSSSTWPSQSITNCNVAHPASSTKVDKKI
VII ARTZAPSVFPLABSCVDTSGSMTTLGCLVKGYFPEPVTVKWNSGAL TSGVHLYPAVL(S).SG(L,Y,S,L,T,S)MVTVPSSQKK.....ATCNVDHPASGTKVDKIV
VIII ARTTAPSVFPLAASCVDTSGSMMTLGCLVKGYFPEPVTVKWNSGAL TSGVHTFPAVLQ.SGLYSLTSMVTVPSSQKK.....ATCNVAHPASSTKVDKTV
IX SGTKAPSVFPLAPCCGDTPSSTVTLGCLVKGYLPEPVTVTWNSGTL TDGVRTFPSVRQSSGLYSVPSTVSVSZP(P,S).......TCBVAH.ATBTKVDKTV
HINGE REGION
Human γ1 (Eu,He) EPKSCDKTHTCPP..CPAPEL
Human γ1 (Nie) EPKSCDKTHTCPP..CPAPQL
Human γ2 (Til) ERKCC...VECPP..CPAPPV
Mouse γ1 (MOPC 21) VPRDC....GCKPCICTVPE.
Mouse γ2a (MOPC 173) EPRGPT.IKPCCPPKCPAPNL
Guinea Pig γ1 QSWGHT.CPPCIP..CGAPZL
Guinea Pig γ2 EPIRTPZPBPCTCPKCPPPEN
Rabbit (Pool) APSTC.....SKPM.CPPPEL
CHII DOMAIN
I LGGPSVFLFPPKPKDTLMISRTPEVTCVVVDVSHEDPQVKFNWYVDGVQV HNAKTKPREQQYBSTYRVVSVLTVLHQNWLDGKEYKCKVSNKALPAPIEKTISKAKG
II LGGPSVFLFPPKPKDTLMISRTPEVTCVVVDVSHEDPEVKFNWYVDGVQV HNAKTKPREEQYBSTYRVVSVLTVLHQDWLDGKEYKCKVSNKALPAPIEKTISKAKG
III .AGPSVFLFPPKPKDTLMISRTPEVTCVVVDVSHEDPEVQFNWYVDGVEV HNAKTKPREEQFNSTFRVVSVLTVVHQDWLDGKEYKCKVSNKGLPAPIEKTISKTKG
IV (←*→)ISRTPEVTCVVVDVSHEDPEVQFNWYVDGVEV HNAKTKPR(Z,Z,Z,F,B,S,T)FRVVSVLTVVHQDWLDGKQYKCKVSNKGLPAPIEKTISKTKG
V ..VSSVFIFPPKPKDTLLITVTPKVTCVVVDISKDDPEVQFSWFVDNVEV HTAQTQPREEQFNSTFRVVSALPIMHQDWLNGKEFKCRVNSAAFPAPIEKTISKTKG
VI LGGPSVFIFPPVKIKNPLMISLSPIVTCVVVDVSEDDPDVQISWFVDNVEV HQAQTTHTRQNYBSTLRVVSALPIQHQNWMSGKEFKCKVNNKDLPAPIERTISKPKG
VII LGGPSVFIFPPKPKBTLMISLTPKATCVVVDVSQEDLEVQFTWYM(N,G)KLM BTAZABVLZZZFNGTFRVVSALTIZHB(←*→)ITRTISKAKG
VIII LGGPSVFIFPPKPKDTLMISLTPRVTCVVVDVSQDEPEVQFTWFVDNKPV GNAETKPRVEQYNTTFRVESVLPIQHQDWLRGKEFKCKVYNKALPAPIEKTISKTKG
IX LGGPSVFIFPP(K,P)KDTLMISRTPEVTCVV.DVSZB(B,P,Z,V,Z)FTWYIDBZQV RTARPPLREQQFDSTIRVVSTLPIAHEDWLRGKEFKCKVHDKALPAPIEKTISKARG
CHIII DOMAIN
I QPREPQVYTLPPSREEMTKNQVSLTCLVKGFYPSDIAVEWESNDGEPEN ..YKTTPPVLDSDGSFFLYSKLTVDKSRWQQGNVFSCSVMHEALHNHYTQKSLSLSPG
II QPREPQVYTLPPSRDELTKNQVSLTCLVKGFYPSDIAVEWESBGEPZDN ..YKTTPPVLDSDGSFFLYSKLTVDKSRWQQGNVFSCSVMHEALHNHYTQKSLSLSPG
III QPREPQVYTLPPSREEMTKNQVSLTCLVKGFYPSDIAVEWESNGQPENN ..YKTTPPMLDSDGSFFLYSKLTVDKSRWQQGNVFSCSVMHEALHNHYTQKSLSLSPG
IV QPREPQVYTLPPSRZEMTKNQVSLTCLVKGFYPSDIAVEWESNGQPENN ..YKTTPPMLDSDGSFFLYSKLTVDKSRWQQGNVFSCSVMHEALHNHYTQKSLSLSPG
V KPRAPQVYTIPPPKEQMAKDKVSLTCMITDFFPEDITVEWQSDGQAPEN ..YKNTQPIMDTDGSYFVYSKLNVQKSNWEAGNTFTCSVLHEGLHNHHTEKSLSHSPG
VI SVRAPQVYVLPPP.ZZMTKKEVTLTCMVTNFMPEDIYVEWTNNGKTELN ..YKNTQPVLDSDGSYFMYSKLRVEKKNWVERNSYSCSVVHQGLHNH(V,S)TKSFSRTPG
VII PPRIPZVYLLPPPRBZLSKKKVSLTCMITGFYPADINVEWDSSE..PSD ..YKNTPPVFDSDETFFLYSRLKVDTNAWNDGESFTCSVMHEALPNHVIQKSISRSPG
VIII APRMPDVYTLPPSRDELSKSKVSVTCLIINFFPADIHVEWASNRVPVSE KEYKNTPPIEDADGSYFLYSKLTVDKSAWDQGTVYTCSVMHEALHNHVTQKAISRSPG
IX QPLEPKVYTMGPPREQLSSRSVSLTCMIDGFYPSDISVGWEKDGKAEDD ..YKTTPAVLDSDGSWFLYSKLSVPTSEWQRGDVFTCSVMHEALHNHYTQKSISRSPG

Fig. 7. Comparisons of C-region amino acid sequences of various mammalian γ chains. Residues were numbered according to the sequence of protein Eu. Positions where all the γ chains compared have an identical amino acid are enclosed in boxes. Periods at a residue position indicate sequence gaps that were introduced to assure maximum homology. The proteins are (I) human γ1 chain Eu and He, (II) human γ1 chain Nie, (III) human γ2 chain Til, (IV) human γ2 chain Zie, (V) mouse γ1 chain MOPC-21, (VI) mouse γ2a chain MOPC-173, (VII) guinea pig γ1 chain, (VIII) guinea pig γ2 chain, and (IX) rabbit γ chain. Sequences in parentheses were not determined. (From A. C. Wang, E. Tung, and H. H. Fudenberg, *J. Immunol.* 125, 1048, 1980).

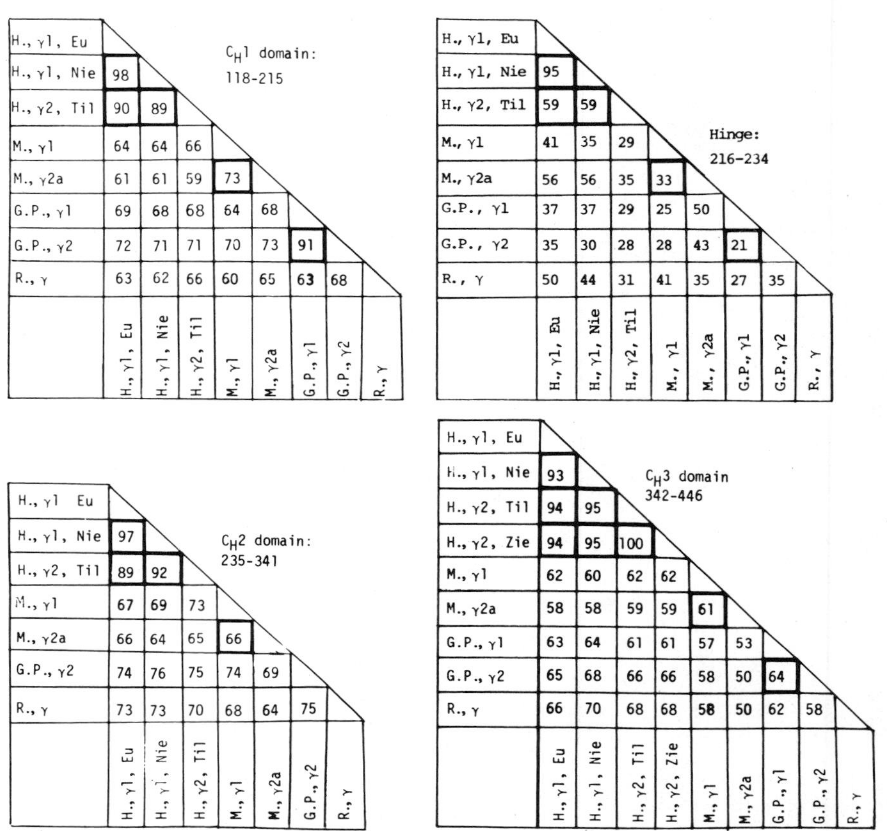

C_H1 domain: 118-215

	H., γ1, Eu	H., γ1, Nie	H., γ2, Til	M., γ1	M., γ2a	G.P., γ1	G.P., γ2	R., γ
H., γ1, Eu								
H., γ1, Nie	98							
H., γ2, Til	90	89						
M., γ1	64	64	66					
M., γ2a	61	61	59	73				
G.P., γ1	69	68	68	64	68			
G.P., γ2	72	71	71	70	73	91		
R., γ	63	62	66	60	65	63	68	

Hinge: 216-234

	H., γ1, Eu	H., γ1, Nie	H., γ2, Til	M., γ1	M., γ2a	G.P., γ1	G.P., γ2	R., γ
H., γ1, Eu								
H., γ1, Nie	95							
H., γ2, Til	59	59						
M., γ1	41	35	29					
M., γ2a	56	56	35	33				
G.P., γ1	37	37	29	25	50			
G.P., γ2	35	30	28	28	43	21		
R., γ	50	44	31	41	35	27	35	

C_H2 domain: 235-341

	H., γ1, Eu	H., γ1, Nie	H., γ2, Til	M., γ1	M., γ2a	G.P., γ2	R., γ
H., γ1 Eu							
H., γ1, Nie	97						
H., γ2, Til	89	92					
M., γ1	67	69	73				
M., γ2a	66	64	65	66			
G.P., γ2	74	76	75	74	69		
R., γ	73	73	70	68	64	75	

C_H3 domain 342-446

	H., γ1, Eu	H., γ1, Nie	H., γ2, Til	H., γ2, Zie	M., γ1	M., γ2a	G.P., γ1	G.P., γ2	R., γ
H., γ1, Eu									
H., γ1, Nie	93								
H., γ2, Til	94	95							
H., γ2, Zie	94	95	100						
M., γ1	62	60	62	62					
M., γ2a	58	58	59	59	61				
G.P., γ1	63	64	61	61	57	53			
G.P., γ2	65	68	66	66	58	50	64		
R., γ	66	70	68	68	58	50	62	58	

Fig. 8. Percent homology among C-region amino acid sequences of various mammalian γ chains. Percentages for intraspecies comparisons are enclosed in thicker boxes. Positions involving B versus D or N and those involving Z versus E or Q were not included in the percentage calculations. (From A. C. Wang, E. Tung, and H. H. Fudenberg, *J. Immunol.* 125, 1048, 1980.)

of DNA may be necessary for γ-chain synthesis, bringing the C_H1 exons from one germ-line DNA segment to the neighborhood of C_H2 and C_H3 exons at another noncontigous germ-line DNA segment during B-cell differentiation.

IV. REGULATORY CONTROL OF IMMUNOGLOBULIN SYNTHESIS

Genetic and biochemical studies have shown that Ig molecules have two kinds of allotypes, called simple and complex allotypes. Alternative forms of both kinds of allotypes generally segregate as codominate alleles in pedigree analysis. However, simple allotypes are associated with only one or a few amino acid substitutions, and most of the Ig allotypes belong to this category (Table 1). In contrast, alternative forms of complex allotypes differ by many amino acid residues (over 10%). The rabbit group a V_H allotypes and group b C_K allotypes, and rat C_K allotypes of the R_L series fall into the complex allotype category (Table 1). Preliminary sequence data (unpublished) suggest that the human V_H allotype, the first reported allotype of the human heavy-chain variable region (Wang et al., 1978; Pandey et al., 1980), also belongs to this category. Several models have been postulated to explain the behavior of the complex allotypes on a true allelic basis, but none is satisfactory. The legitimacy of these allotypes has long been questioned (Wang, 1974). They may actually be multiple closely linked isotypes, because (1) a case in which one rabbit produced more than two group a and more than two group b allotypes has been reported (Stronsberg, 1977), and (2) many rabbits of different genetic backgrounds are able to produce low levels of group a allotypes not predicted by their genotypes (Mudgett et al., 1975).

One model for the genetic control of complex allotypes is presented in Fig. 9. According to this model, complex allotypes are synthesized by clusters of closely linked genes which have arisen by gene duplication and divergence. But the expression of these gene clusters is controlled by different allelic variants of a hypothetical regulatory gene.

V. ANTIBODY DIVERSITY

The genetic control of antibody diversity, though vigorously pursued by experimental biologists for many years, is still not fully understood. Many hypotheses have been formulated, and the major ones are listed below:

1. *Germ Line*. This is the most conservative hypothesis, involving the least number of *ad hoc* assumptions. According to this hypothesis, there is a gene for each V region inherited in the genome. V-region genes arose by normal processes of molecular evolution, i.e., by gene duplication followed by mutation and natural selection. The ability to produce antibody toward a given antigen is passed on from generation to generation. The total number of V-region genes would be over 1,000 for each of the V_H and V_L regions (Hood and Talmage, 1970).

Table 1. Amino acid interchanges probably associated with Allotypic differences in immunoglobulins.

Chain	Residue Number	Amino Acid Residue	Associated Allotype
Human γ1	356,358	Asp, Leu	G1m(1)
		Glu, Met	G1m(1-)
Human γ1	431	Gly	G1m(2)
		Ala	G1m(2-)
Human γ1	214	Arg	G1m(3)
		Lys	G1m(3-)
Human γ3	296,436	Phe, Phe	G3m(5)
		Tyr, Tyr	G3m(21)
Human γ2	212,221	Ser, Arg	A2m(2)
		Pro, Pro	A2m(1)
Human κ	153,191	Val, Leu	Km(1)
	153,191	Ala, Leu	Km(1,2)
	153,191	Ala, Val	Km(3)
Human H	V Region	Multiple Differences	Hv(1), Hv(1-)
Mouse α	C_H1 Domain	Bond Between H & L Chains	In NZB Mice
		No Bond	In BALB/c Mice
Mouse γ2a	135	Ser	In C_3H Mice
		Thr	In BALB/c Mice
Mouse κ	C Region	Multiple Differences	b4, b5, b6, b9
Rat κ	C Region	Multiple Differences	RL-1a, RL-1b
Rabbit κ	225	Met	d11
		Thr	d12
Rabbit γ	309	Thr	c14
		Ala	c15
Rabbit H	V Region	Multiple Differences	a1, a2, a3

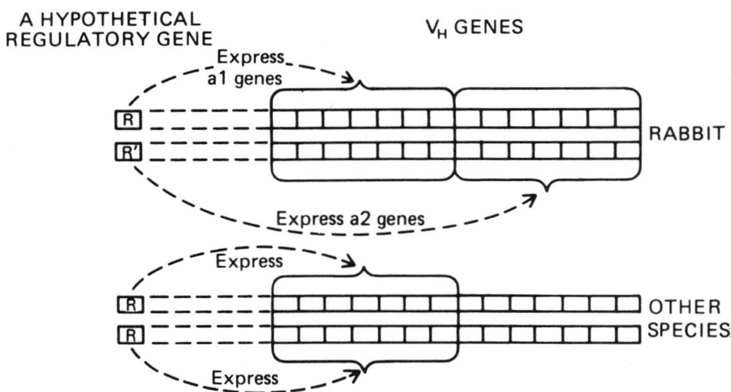

Fig. 9. A hypothetical explanation of complex (or latent) allotypes, using the rabbit group a (Aa or allotype a) V_H allotypes as an example. R and R' represent alternative alleles of a hypothetical regulatory gene that controls the expression of families of closely linked V_H genes. (From A. C. Wang, *Proc. Fed. Biochem. Symp.* 36, 19-31, 1974).

2. *Somatic Mutation.* This hypothesis was proposed independently by Burnet (1959) and Lederberg (1959). It suggests that there are only a few V-region genes (perhaps one for each subgroup). These genes are highly mutable and become diversified in somatic cells, yielding differentiated clones of B lymphocytes producing antibodies that differ in the V region. Hypervariable regions correspond to "hot spots". Natural selection operates on individual cells rather than on individual organisms.

3. *V Genes Encoding Autoantibodies to Histocompatibility Antigens.* This is a modified version of the somatic mutation hypothesis. Jerne (1971) proposed that V-region genes in an animal code for antibodies directed against histocompatibility antigens of the same species. There is a strong selection against lymphocytes whose V-region genes have not mutated, because lymphocytes synthesizing antibodies to self constituents of the animal would be destroyed or at least suppressed. There are few V-region genes in the genome.

4. *DNA Repair.* This is another modified version of the somatic mutation hypothesis. Brenner and Milstein (1966) postulated that partial degradation of V-region genes and the incorporation of errors during the repairing process could be responsible for the generation of antibody diversity. The actions of a cleavage enzyme and exonuclease remove portions of a gene; this is followed by repair, with errors, to produce a new V-region gene. Enzyme called terminal deoxynucleotidyl transferase, which has been found in the thymus of several species, can add nucleotides to the 3'-OH end of DNA. The nucleotides added do not need to be complementary to the second strand.

5. *Somatic Recombination.* This hypothesis was proposed by Smithies (1967), who suggested that there are a moderate number (a few per subgroup) of V-region genes, and that recombinations among V-region genes within the same subgroup produce new V-region genes in somatic cells. V-region genes are selected in evolution for their ability to generate recombinants rather than their ability to determine V regions of a particular antigen-binding specificity.

6. *Predetermined Premutation of Germ-Line Genes.* Kliman et al. (1976) suggested that there are hundreds of V-region genes inherited in the genome. These genes undergo a predetermined path of mutation in somatic cells to give rise to thousands of V-region gene variants, enough to code for more than ten million antibodies. These mutations are not antigen-driven. The newborn repertoire is 10^4 "clonotypes", and the adult repertoire is 10^7 clonotypes.

7. *Mini-gene Combination.* Based on analysis of V-region amino acid sequences reported in the literature, Wu and Kabat (1970) suggested that each V region is encoded by many "mini-genes", in the sense that each hypervariable segment and each framework segment is encoded by a separate small gene. Assuming that there are 5 to 10 variant forms of mini-genes for each of the hypervariable and framework segments, and assuming that any of the mini-gene variants for a given segment can combine freely with any member of another segment, there could be up to a million possible combinations for the V_H as well as the V_L region. The J segment has been shown to behave as a mini-gene, and the D segment is another likely candidate, but there has been no strong evidence to support the existence of other mini-genes.

During the 1940s and 1950s many immunologists believed that antigens could instruct the organism to produce complementary antibodies by serving as templates around which the antibodies might fold. This hypothesis was rejected in the early 1960s, when it was shown that antibodies were able to retain their initial antigen-binding specificities following denaturation and subsequent renaturation in the absence of the respective antigen. Subsequent attempts to explain antibody diversity have focused on the origin of primary structure variation among the V_H and V_L regions of polypeptide chains.

The classical germ-line (Hood and Talmage, 1970) and somatic mutation (Cohn, 1971) hypotheses dominated in the 1960s and the first half of the 1970s, but were later shown to be unacceptable. The germ-line hypothesis could not explain how during evolution certain V-region genes that encode antibodies against synthetic haptens, which have never existed in the past, could be selected. On the other hand, the discovery that certain idiotypes are inherited in animals (Eichmann, 1975) silenced any argument based strictly on somatic mechanisms, because mutations and recombinations are in general random events.

The mini-gene combinatory hypothesis (Kabat, 1980; Kindt and Capra, 1978) emerged as the leading theory in the mid-1970s. However, the finding that nucleotides encoding hypervariable regions and those encoding framework regions are in one continuous piece in germ cells (Davis et al., 1980) and embryonic cells (Sakano et al., 1980) took some steam out of the mini-gene combination hypothesis.

DNA-RNA hybridization, gene cloning, and DNA sequence analyses indicate that the mouse has hundreds of germ-line V_κ genes (Seidman et al., 1978). During the maturation of B cells, one of these V_κ genes is translocated to one of five J_κ segments by site-specific recombination (Seidman et al., 1979). The site of recombination coincides with the third hypervariable region. Assuming that any of the V_κ genes can associate with any of the J_κ segments, and that new codons can be generated at the recombination site by varying the precise position of the translocation within the combination site, then thousands or more κ chains can be produced in the mouse.

The diversity of V_H regions is further amplified by the presence of the D segments. The V_H exons do not join directly to the J segments but rather to a third set of DNA segments, designated the D segments, which in turn join to the J_H segments. It is noteworthy that the D segments, which vary considerably in length, also correspond to the third hypervariable region. With hundreds of V_H genes and moderate numbers of D and J_H segments, it would not be difficult for a mouse to produce thousands or more V_H regions. It is conceivable that further somatic diversification in the V region may occur after the V-J translocation. However, recombination between the V and J regions (or between V and D as well as between D and J in the H chain) should generate enough V-region variability to account for antibody diversity even without somatic mutation or recombination.

The author thanks Dr. I. Y. Wang for comments and criticisms, Mr. Charles L. Smith for editorial assistance and Ms. Linda Westerberg and Nancy Butler for secretarial assistance. This work was supported in part by a National Science Foundation Research Grant (PCM-82-01751) and by an MUSC Biomedical Research Grant (CR-10). This is publication No. 604 from the Department of Basic and Clinical Immunology and Microbiology, Medical University of South Carolina.

VI. REFERENCES

Arnaud, P., Wang, A. C., Gianazza, E., Wang, I. Y., Lasne, Y., Creyssel, R., and Fudenberg, H. H. (1981). Gamma heavy chain disease protein Cha: Immunological and structural studies. *Mol. Immunol.*, 18, 379-384.

Brenner, S. and Milstein, C. (1966). Origin of antibody variation. *Nature*, 211, 242-243.

Brunhouse, R. and Cebra, J. J. (1979). Isotypes of IgG: comparison of the primary structures of three pairs of isotypes which differ in their ability to activate complement. *Mol. Immunol.*, 16, 907-917.

Burnet, F. M. (1959). The clonal selection theory of acquired immunity. *Cambridge University Press, London.*

Cohn, M. (1971). The take home lesson. *Ann. N.Y. Acad. Sci.*, 190, 529-584.

Coleclough, C., Cooper, D., and Perry, R. P. (1980). Rearrangement of immunoglobulin heavy chain genes during B-lymphocyte development as revealed by studies of mouse plasmacytoma cells. *Proc. Nat. Acad. Sci. U.S.A.*, 77, 1422-1426.

Colomb, M. and Porter, R. R. (1975). Characterization of a plasmin-digest fragment of rabbit immunoglobulin gamma that binds antigen and complement. *Biochem. J.*, 145, 177-183.

Connell, G. E., Parr, D. M., and Hofmann, T. (1979). The amino acid sequence of the three heavy chain constant region domains of a human IgG2 myeloma protein. *Can. J. Biochem.*, 57, 758-767.

Cory, S., Jackson, J., and Adams, J. M. (1980). Deletions in the constant region locus can account for switches in immunoglobulin heavy chain expression. *Nature*, 285, 450-456.

Couderc, J., Bleux, C., Ventura, M., and Liacopoulos, P. (1979). Single mouse cells producing two antibody molecules and giving rise to antigen-driven intraclonal variation after immunization with two unrelated antigens. *J. Immunol.*, 123, 173-181.

Cunningham, B. A., Gottlieb, P. D., Pflumm, N. M., and Edelman, G. M. (1971). Immunoglobulin structure: Diversity, gene duplication and domains. *Prog. Immunol.*, 1, 3-24.

Davis, M. M., Kim, S. K., and Hood, L. E. (1980). DNA sequences mediating class switching in α-immunoglobulins. *Science*, 209, 1360-1365.

Early, P. W., Davis, M. M., Kaback, D. B., Davidson, N., and Hood, L. (1979). Immunoglobulin heavy chain organication in mice: analysis of a myeloma genomic clone containing variable and alpha constant regions. *Proc. Nat. Acad. Sci. U.S.A.*, 76, 857-861.

Edelman, G. M., Cunningham, Gall, E. W., Gottlieb, P. D., and Waxdal, N. (1969). The covalent structure of an entire G1 immunoglobulin molecule. *Proc. Nat. Acad. Sci. U.S.A.*, 63, 78-85.

Eichmann, K. (1975). Genetic control of antibody specificity in the mouse. *Immunogenetics*, 2, 491-506.

Fougereau, M., Bourgois, A., dePreval, C., Rocca-Serra, J., and Schiff, C. (1976). The complete sequence of the murine monoclonal immunoglobulin MOPC 173 (IgG2a): Genetic implications. *Ann. Immunol. Inst. Pasteur (Paris)*, 127C, 607-631.

Franklin, E. C. and Frangione, B. (1975). Structural variants of human and murine immunoglobulins. *Contemp. Top. Mol. Immunol.*, 4, 89-125.

Fruchter, R. G., Jackson, S. A., Mole, L. E., and Porter, R. R. (1970). Sequence studies of the Fd section of the heavy chain of rabbit immunoglobulin G. *Biochem. J.*, 116, 249-259.

Fudenberg, H. H., Pick, J. R. L., Wang, A. C., and Douglas, S. D. (1978). *Basic Immunogenetics* (2nd Ed.), *Oxford University Press, New York.*

Gally, J. A. and Edelman, G. M. (1972). The genetic control of immunoglobulin synthesis. *Ann. Rev. Genet.*, 6, 1-46.

Hill, R. L., Delaney, R., Fello, R. E., and Lebowitz, J. D. (1966). The evolutionary origins of immunoglobulins. *Proc. Nat. Acad. Sci. U.S.A.*, 56, 1762-1769.

Honjo, T. and Kataoka, T. (1978). Organization of immunoglobulin heavy chain genes and allelic deletion model. *Proc. Nat. Acad. Sci. U.S.A.*, 75, 2140-2144.

Hood, L. (1972). Two genes, one polypeptide chain--facts or fiction? *Fed. Proc.*, 31, 177-187.

Hood, L., Campbell, J. H., and Elgin, S. C. R. (1975). The organization, expression and evolution of antibody genes and other multigene families. *Ann. Rev. Genet.*, 9, 305-353.

Hood, L. and Talmage, D. (1970). Mechanism of antibody diversity: Germ line basis for variability. *Science*, 168, 325-334.

Hozumi, N. and Tonegawa, S. (1976). Evidence for somatic rearrangement of immunoglobulin genes coding for variable and constant region. *Proc. Nat. Acad. Sci. U.S.A.*, 73, 3628-3632.

Jerne, N. K. (1971). The somatic generation of immune recognition. *Eur. J. Immunol.*, 1, 1-9.

Joho, R. and Weissman, I. L. (1980). V-J joining of immunoglobulin K genes occurs on one homologous chromosome. *Nature*, 284, 179-181.

Kabat, E. A. (1980). Origin of antibody complementary and specificity-hypervariable regions and the minigene hypothesis. *J. Immunol.*, 125, 961-969.

Kataoka, T., Kawakami, T., Takahashi, N., and Honjo, T. (1980). Rearrangement of immunoblobulin γ1 chain gene and mechanism for heavy chain class switch. *Proc. Nat. Acad. Sci. U.S.A.*, 77, 919-923.

Kindt, T. J. and Capra, D. (1978). Gene-insertion theories of antibody diversity; A re-evaluation. *Immunogenetics*, 6, 309-321.

Kilinman, N. R., Sigal, N. H., Metcalf, E. S., Pierce, S. K., and Gerhart, P. J. (1976). The interplay of evolution and environment in B-cell diversification. *Cold Spring Harbor Symp. Quant. Biol.*, 41, 165-173.

Kunkel, H. G., Natvig, J. B., and Joslin, F. G. (1969). A "lepore type of hybrid γ-globulin. *Proc. Nat. Acad. Sci. U.S.A.*, 62, 144-149.

Lederberg, S. (1959). Gene and antibodies. *Science*, 129, 1649-1653.

Liu, C. P., Tucker, P. W., Mushinski, J. F., and Blattner, F. R. (1980). Mapping of heavy chain genes for mouse immunoglobulin M and D. *Science*, 209, 1348-1353.

Marchalonis, J. J. (1980). Molecular interactions and recognition specificity of surface receptor. *Contemp. Top. Immunobiol.*, 9, 255-288.

Milstein, C., Adetugbo, K., Cowan, N. J., and Secher, D. C. (1974). Clonal variants of myeloma cells. *Prog. Immunol.*, 2, 157-168.

Moore, K. W., Rogers, J., Hunkapiller, T., Early, P., Nottenburg, C., Weissman, I., Bazin, H., Wall, R., and Hood, L. E. (1981). Expression of IgD may use both DNA rearrangement and RNA splicing mechanisms. *Proc. Natl. Acad. Sci. (USA)*, 78, 1800-1804.

Mudgett, M., Fraser, B. A., and Kindt, T. J. (1975). Nonallelic behavior of rabbit variable-region allotypes. *J. Exp. Med.*, 141, 1448-1452.

Natvig, J. B., Michaelson, T. E., Gedde-Dahl, T. Jr., and Fischer, R. (1974). IgG1 subclass protein with the genetic Gm markers Gm (Z + non-a +) another example of intrageneic hybridization among IgG subclass genes. *J. Immunogenet.*, 1, 33-41.

Nottenberg, C. and Weissman, I. L. (1981). Cu gene rearrangement of mouse immunoglobulin genes in normal B cells occurs on both the expressed and nonexpressed chromosomes. *Proc. Nat. Acad. Sci. U.S.A.*, 78, 484-488.

Pandey, J. P., Tung, E., Mathur, S., Namboodiri, K. K., Wang, A. C., Fudenberg, H. H., Blattner, W. A., Elston, R. C., and Hames, C. G. (1980). Linkage relationship between variable and constant region allotypic determinants of human immunoglobulin heavy chains. *Nature*, 286, 404-407.

Ponstingl, H and Hilschmann, N. (1972). Die Primarstrucktur eines monoklonalen IgG1 immunoglobulin (myeloprotein Nie). II. Aminosauresequenz des konstanten Teils der H-Kette, zuordnung genetischer Faktoren. *Hoppe-Seyler's Z. Physiol. Chem.*, 353, 1369-1372.

Rabbitts, T. H., Forster, A., Dunnick, W., and Bentley, D. L. (1980). The role of gene deletion in the immunoglobulin heavy chain switch. *Nature*, 283, 351-356.

Ravetch, J. V., Kirsch, I. R., and Leder, P. (1980). Evolutionary approach to the question of immunoglobulin heavy chain switching: Evidence from cloned human and mouse genes. *Proc. Nat. Acad. Sci. U.S.A.*, 77, 6734-6738.

Sakano, H., Huffi, K., Heinrich, G., and Tonegawa, S. (1979a). Sequences at the somatic recombination sites of immunoglobulin light-chain genes. *Nature*, 280, 288-294.

Sakano, H., Rogers, J. H., Huffi, K., Brack, C., Traunecker, A., Maki, R., Wall, R., and Tonegawa, S. (1979b). Domains and the hinge region of an immunoglobulin heavy chain are encoded in separate DNA segments. *Nature*, 277, 627-633.

Sakano, H., Maki, R., Kurosawa, Y., Roeder, W., and Tonegawa, S. (1980). Two types of somatic recombination are necessary for the generation of complete immunoglobulin heavy-chain genes. *Nature*, 280, 676-683.

Seidman, J. G., Leder, A., Edgell, M. H., Polsky, F., Tilghman, S. M., Tiemeier, D. C., and Leder, P. (1978). Multiple related immunoglobulin variable-region genes identified by cloning and sequence analysis. *Proc. Nat. Acad. Sci. U.S.A.*, 75, 3881-3885.

Seidman, J. G., Max, E. E., and Leder, P. (1979). A κ-immunoglobulin gene is formed by site-specific recombination without further somatic mutation. *Nature*, 280, 370-375.

Singer, S. J. and Doolittle, R. F. (1966). Antibody active site and immunoglobulin molecule. *Science*, 153, 13-25.

Smithies, O. (1967). The genetic basis of antibody variability. *Cold Spring Harbor Symp. Quant. Biol.*, 32, 161-168.

Steinberg, A. G., Terry, W. D., and Morrell, A. R. (1970). Human allotype and genetic dogma. *Prot. Biol. Fluids*, 17, 111-116.

Strosberg, A. D. (1977). Multiple expression of rabbit allotypes: The tip of the iceberg? *Immunogenetics*, 4, 449-513.

Tsuzukida, Y., Wang, C. C., and Putman, F. W. (1979). Structure of the A2m(1) allotype of human IgA: A recombinant molecule. *Proc. Nat. Acad. Sci. U.S.A.*, 76, 1104-1108.

Tung, E., Kuan, T. K., Litman, G. W., and Wang, A. C. (1981). Three monoclonal immunoglobulins in one patient. I. Isolation and characterization. *Immunology* (Submitted).

Wabl, M. R., Johnson, J. P., Haas, I. G., Tenkhoff, M., Meo, T., and Inan, R. (1980). Simultaneous expression of mouse immunoglobulins M and D is determined by the same homology of chromosome 12. *Proc. Nat. Acad. Sci. U.S.A.*, 77, 6793-6796.

Wang, A. C. (1974). Gene expression and evolution: Evidence for differential expression of immunoglobulin variable region genes in different species. *Proc. Fed. Biochem. Symp.*, 36, 19-31.

Wang, A. C. (1978). Evidence for and significance of two genes for one polypeptide chains. In *Immunoglobulins* (G. W. Litman and R. A. Good, eds.), *Plenum, New York*, pp. 229-255.

Wang, A. C. and Wang, I. Y. (1979). Immunoglobulin structure and genetics. *Rev. Inst. Pasteur (Lyon)*, 12, 325-345.

Wang, A. C., Arnaud, P., Fudenberg, H. H., and Creyssel, R. (1978). Monoclonal IgM cryoglobulinemia associated with gamma-3 heavy chain disease: Immunochemical and biochemical studies. *Eur. J. Immunol.*, 8, 375-379.

Wang, A. C., Mathur, S., Pandey, J., Siegal, F. P., Middaugh, C. R., and Litman, G. W. (1978). Hv(1), a variable-region genetic markers of human immunoglobulin heavy chains. *Science*, 200, 327-329.

Wang, A. C., Tung, E., and Fudenberg, H. H. (1980). The primary structure of a human IgM2 heavy chain: Genetic, evolutionary, and functional implications. *J. Immunol.*, 125, 1048-1054.

Wang, A. C., Wang, I. Y., and Fudenberg, H. H. (1977). Immunoglobulin structure and genetics; Identity between variable regions of a μ and a γ2 chain. *J. Biol. Chem.*, 252, 7192-7199.

Wang, A. C., Wang, I. Y., McCormick, J. N., and Fudenberg, H. H. (1969). The identity of light chains of monoclonal IgM and monoclonal IgM in one patient. *Immunochemistry*, 6, 451-459.

Wang, A. C., Wilson, S. K., Hopper, J. F., Fudenberg, H. H., and Nisonoff, A. (1970). Evidence for control of synthesis of the variable regions of the heavy chains of immunoglobulins M by the same gene. *Proc. Nat. Acad. Sci. U.S.A.*, 66, 337-343.

Woodland, R. T. and Cantor, H. (1980). V_H gene products allow specific communication among immunologic cell sets. *Contemp. Top. Immunobiol.*, 11, 227-244.

Wu, T. T. and Kabat, E. A. (1970). An analysis of the sequences of the variable regions of Bence-Jones proteins and myeloma light chains and their implications for antibody complementary. *J. Exp. Med.*, 132, 211-250.

Yaoita, Y. and Honjo, T. (1980). Deletion of immunoglobulin heavy chain genes from expressed allelic chromosome. *Nature*, 286, 850-853.

THE HLA SUPERGENE AND DISEASE[1]

Janardan P. Pandey and H. Hugh Fudenberg

Department of Basic and Clinical Immunology
and Microbiology
Medical University of South Carolina
171 Ashley Avenue
Charleston, South Carolina 29425

[1]Publication No. 524 from the Department of Basic and Clinical Immunology and Microbiology, Medical University of South Carolina. Research supported in part by USPHS Grant CA-25746.

I. INTRODUCTION

A "supergene" is a group of linked genes usually inherited as a unit. If co-adapted, such genes (which are not necessarily functionally related) may cooperatively produce some adaptive characteristic (Darlington and Mather, 1969; Rieger et al., 1976). Ceppilini first termed the human major histocompatability complex, HLA, a supergene. It includes many components of widely diverse function and structure. By both family and somatic cell hybridization studies this complex has been mapped on the short arm of chromosome 6. To date, five gene loci for human leukocyte-associated antigens (HLA) have been designated: HLA-A, HLA-B, HLA-C, HLA-D, and HLA-DR; the DR locus has been further divided into three subloci (Markert and Cresswell, 1980). These genes are closely linked (crossing over about 0.8%). HLA antigens A, B, and C are found on the surface of all nucleated cells of the body, whereas D and DR have a more restricted tissue distribution. In addition to the HLA loci, this complex also includes genes controlling the levels of complement components C2, C4, and C8, and a number of enzyme and red cell polymorphisms.

The HLA system is the most polymorphic system known in man. A complete list of currently recognized HLA specificities is given in Table 1. A large number of different determinants have been recognized at each locus, including at the time of this writing 20 for A, 42 for B, 8 for C, 12 for D, and 10 for DR. In combination, these antigens can generate more than 300 million genetically different individuals (Bodmer, 1980).

With random mating, and given enough time in evolution, population genetics theory predicts that the alleles of one locus will be in random association with the alleles of other loci (linkage equilibrium). Despite the fact that crossing-over is known to occur between HLA loci, some HLA alleles (haplotypes) tend to occur together on the same chromosome significantly more often than expected according to their individual gene frequencies, a phenomenon termed "linkage disequilibrium." Well-known examples of linkage disequilibrium in the HLA system are A1-B8 and A3-B7 in Caucasians.

The HLA-A, -B, and -C antigens are composed of two noncovalently linked polypeptide chains; a light chain identical to β_2-microglobulin (molecular weight 11,600) coded by a gene on chromosome 15 and a heavy chain (molecular weight 44,000) carrying the serologically detected antigens coded by gene(s) in the HLA complex on chromosome 6. No allotypic variation in β_2-microglobulin has been reported in man, but in mice a recent report suggests that it is polymorphic (Robinson et al., 1981). The HLA-DR antigens have also been shown to be composed of two polypeptide chains of

Table 1. Complete listing of recognized HLA specificities.*

HLA-A	HLA-B	HLA-C	HLA-D	HLA-DR
HLA-A1	HLA-B5	HLA-Cw1	HLA-Dw1	HLA-DR1
HLA-A2	HLA-B7	HLA-Cw2	HLA-Dw2	HLA-DR2
HLA-A3	HLA-B8	HLA-Cw3	HLA-Dw3	HLA-DR3
HLA-A9	HLA-B12	HLA-Cw4	HLA-Dw4	HLA-DR4
HLA-A10	HLA-B13	HLA-Cw5	HLA-Dw5	HLA-DR5
HLA-A11	HLA-B14	HLA-Cw6	HLA-Dw6	HLA-DRw6
HLA-Aw19	HLA-B15	HLA-Cw7	HLA-Dw7	HLA-DR7
HLA-Aw23(9)	HLA-Bw16	HLA-Cw8	HLA-Dw8	HLA-DRw8
HLA-Aw24(9)	HLA-B17		HLA-Dw9	HLA-DRw9
HLA-A25(10)	HLA-B18		HLA-Dw10	HLA-DRw10
HLA-A26(10)	HLA-Bw21		HLA-Dw11	
HLA-A28	HLA-Bw22		HLA-Dw12	
HLA-A29	HLA-B27			
HLA-Aw30	HLA-Bw35			
HLA-Aw31	HLA-B37			
HLA-Aw32	HLA-Bw38(w16)			
HLA-Aw33	HLA-Bw39(w16)			
HLA-Aw34	HLA-B40			
HLA-Aw36	HLA-Bw41			
HLA-Aw43	HLA-Bw42			
	HLA-Bw44(12)			
	HLA-Bw45(12)			
	HLA-Bw46			
	HLA-Bw47			
	HLA-Bw48			
	HLA-Bw49(w21)			
	HLA-Bw50(w21)			
	HLA-Bw51(5)			
	HLA-Bw52(5)			
	HLA-Bw53			
	HLA-Bw54(w22)			
	HLA-Bw55(w22)			
	HLA-Bw56(w22)			
	HLA-Bw57(17)			
	HLA-Bw58(17)			
	HLA-Bw59			
	HLA-Bw60(40)			
	HLA-Bw61(40)			
	HLA-Bw62(15)			
	HLA-Bw63(15)			
	HLA-Bw4			
	HLA-Bw6			

*From Histocompatibility Testing 1980.

unequal size (molecular weights 35,000 and 27,000); both chains are different from those which constitute the A, B, and C antigens. In contrast to the A, B, and C antigens, where only the heavy chain is polymorphic, in HLA-DR both heavy and light chains are polymorphic (Kaufman et al., 1980; G. B. Ferarra, pers. commun.).

II. HLA AND DISEASE ASSOCIATIONS

A. Basic Considerations

Diseases may be associated and/or linked to HLA. The terms "linkage" and "association" are often confused. An association is demonstrated in population studies when a specific allele or alleles are noted to be found in persons with a given disease more often than in ghe general population. Linkage, on the other hand, means that two gene loci are present on the same chromosome. In a random-mating equilibrium population, zygotic proportions for linked genes and for those assorting independently are identical. Consequently, linkage of genes cannot be detected by examining genotypic or phenotypic proportions in a general population. Any observed associations of traits in the general population may be due to a variety of causes: common environmental influences, multiple effects of the same gene or genes (pleiotropy), subdivision of a population into heterogeneous groups, differential inbreeding, physiological consequences of development, epistasis (Clerget-Darpoux and Bonaiti-Pellié, 1980), or simply bias in sampling procedure. Thus, association does not constitute evidence for genetic linkage (Li, 1976). Population associations between linked traits due to their linkage exist only if there is significant linkage disequilibrium (Bodmer and Payne, 1965).

Statistical methods for measuring association are different from those for detecting linkage. The strength of a population association is usually expressed as relative risk (Woolf, 1955), which is simply a cross-product ratio of the entries in a 2x2 contingency table. Relative risk indicates how many more times a disease is likely to occur in individuals possessing a certain antigen than in those lacking that antigen. Statistical significance of associations can be expressed in terms of P values; however, the number of comparisons made must be taken into account. To avoid deviations due to pure chance, the P values should be multiplied by the number of antigens tested (Grumet et al., 1971). This corrected P value, sometimes called the experiment-wise error rate, is very conservative; Hartley (1955) has recommended 0.1 higher levels for rejection points.

Family studies are usually necessary to detect genetic linkage. The lod score method of Morton (1955), which is generally used to test for linkage, determines the relative odds that a sequence of genotypes among siblings resulted from linked loci with a certain percentage of meiotic crossing over, as compared with the possibility that the sequence occurred by change from unlinked loci. For convenience, the ratio of these two probabilities--recombination fraction $\theta < \frac{1}{2}$ (linkage) compared with $\theta = \frac{1}{2}$ (no linkage) is expressed as its logarithm. The $\log_{10}$ of this relative probability is called the log of odds or the "lod score", and the value of θ which corresponds to the maximum value of the lod score is taken as the estimate of the recombination fraction, θ. A negative value of the lod score argues against linkage; a value greater than 2 is suggestive of linkage; a value greater than 3 is accepted as proof of linkage. Alternative methods developed by Haseman and Elston (1970), Day and Simons (1976), and Suzrez (1978) can also be used to detect linkage of disease susceptibility genes to the polymorphic loci in multiple case families, i.e., families containing two or more affected sibs.

B. The Associations

Over 100 diseases have been studied in regard to HLA associations. Table 2 summarizes some of the associations which have been confirmed by several studies. This information has been extracted from the 1980 Histocompatibility Workshop (Terasaki, 1980) and from Svejgaard et al. (1980). For each disease and the antigen associated, the relative risk of the disease for individuals carrying that antigen is given. For instance, individuals positive for B27 are over 90 times more likely to develop ankylosing spondylitis than are individuals lacking this antigen.

Although Hodgkin disease was the first to be found associated with HLA, subsequent investigations have shown very weak, if any, associations between HLA and various malignancies. There may be several reasons for this lack of association (Dausset, 1977): (1) many more genes may be involved in malignancies than in other disorders, and the major gene may not be in linkage disequilibrium with the HLA complex; (2) if most of the associations between HLA and disease are reflections of immune response genes, one would not expect to find an association between HLA and malignancies, since immune responses are often weak or absent in patients with malignancies.

Associations between various HLA antigens and diseases can be divided into two categories: those present in high frequency in patients studied at the onset of disease and those present in high frequency in patients studied later in the disease course. The former represent susceptibility, whereas the latter probably indicate resistance to death from the disease. For example, in Hodgkin disease, the presence of HLA-B8 is associated with

DISEASE	HLA	RELATIVE RISK
Idiopathic hemochromatosis	A3	4
	B14	5
Behcet's disease	B5	6
Ankylosing spondylitis	B27	90
Reiter's disease	B27	37
Congenital adrenal hyperplasia (21-hydroxylase deficiency)	Bw47	15
Psoriasis	Cw6	13
	DR7	43
Insulin-dependent diabetes mellitus	DR3	3
	DR4	6
	DR3/DR4	33
Chronic active hepatitis	B8	7
	DR3	2
Rheumatoid arthritis		
Adult onset	DR4	3
Juvenile onset	DR5	2
	DR8	2
Multiple sclerosis	B7	4
	DR2	4
Membranous glomerulonephritis	DR3	4
Myasthenia gravis	B8	3
	DR3	3
Celiac disease	DR3	17
	DR7	4
Systemic lupus erythematosus	B8	3
	DR3	6
Graves disease	DR3	4

increased survival time (Falk and Osoba, 1971), whereas Aw19 and/ or B5 are apparently associated with decreased survival (Falk and Osoba, 1977).

Although disease association studies in general have been designed to look for susceptibility to a disease with an increased frequency of a known HLA antigen, a decreased frequency of an antigen would imply some form of resistance to the disease. For instance, in multiple sclerosis A2 and B12 are decreased, and in celiac disease and diabetis mellitus B7 decreased. However, studies showing associations with decreased frequencies of HLA antigens require larger study populations and are more difficult to perform (Braun, 1979).

C. Interactive Effect of HLA and Gm

Some recent studies suggest that HLA and Gm (genetic markers of IgG) somewhat interact both in normal and pathological conditions. For example, Whittingham et al. (1980) have shown an interactive effect of Gm and HLA in immune response to the bacterial antigen monomeric flagellin. Within Gm phenotype subgroups, HLA-B7, B8, and B12 had significant effects on mean log titers of IgG antibody to flagellin. Among subjects with Gm(a-x-b+), for example, those with HLA-B7 tended to be low responders and those with HLA-B8 tended to be high responders. It has been postulated that the interactive effects of Gm and HLA on antibody titer reflect the recognition of determinants on the flagellin molecule by different populations of lymphocytes. Such interactions between Gm and HLA may provide not only a mechanism for increasing the specificity of antibody reactions but also a more flexible means of evolutionary adaptation to newly encountered and potentially harmful infections (Whittingham et al., 1980).

Interaction of HLA and Gm in autoimmune chronic active hepatitis has been shown by the same authors (Whittingham et al., 1981). Compared with the low risk group (i.e., controls) who were HLA-B8 negative and Gm a+x+ positive, the relative risk was 39 times greater in subjects with both Gm a+x+ and HLA-B8, 15 times greater in subjects with HLA-B8 but lacking Gm a+x+, and two times greater in subjects with neither HLA-B8 nor Gm a+x+.

D. Diagnostic and Prognostic Value

Ankylosing spondylitis is the only disease for which HLA typing can sometimes be used as a "diagnostic" test. Since approximately 10% of the individuals with ankylosing spondylitis lack B27 and about 8% of normal population has B27, only about 10% false negative and 8% false positive results should be obtained. Also, the diagnostic value of HLA typing in akylosing spondylitis depends on the *a priori* probability of the disease based on clinical and laboratory findings.

In congenital adrenal hyperplasia due to 21-hydroxylase (21-OH) deficiency, it is possible to HLA type fetal cells and thus make a prenatal diagnosis in offspring of parents who have had a child with 21-OH deficiency. All offspring who have the same HLA haplotype as the affected child will be expected to develop congenital adrenal hyperplasia (Svejgaard et al., 1980).

In some cases, a disease runs a more severe course in pateints with a given antigen than in those lacking it. For instance, D/DR2 positive multiple sclerosis patients have more severe clinical courses than thos lacking these antigens. Other examples are chronic active hepatitis and rheumatoid arthritis (van Rood et al., 1981).

E. Contributions to the Genetics of a Disease

The genetics of diabetes is still an enigma. No clear-cut mode of inheritance has been established. However, HLA typing studies have shown that insulin-dependent and noninsulin-dependent diabetes have different genetic backgrounds, since only the former is associated with HLA and not the latter. Another example in this category is psoriasis: only psoriasis vulgaris and not postular psoriasis is associated with HLA. Moreover, HLA typing has indicated heterogeneity within psoriasis vulgaris, since only the early onset and not the late onset type seems to be HLA associated.

HLA typing has also provided evidence for a common genetically determined defect in different diseases, e.g., B8 associated diseases have been suggested to have a common etiology, and they often occur together in the same individual.

The loci for idiopathic hemochromatosis and for congenital adrenal hyperplasia due to 21-OH deficiency, neither of which involves any known immunological function, have been mapped in the HLA region. HLA studies have shown that idiopathic hemochromatosis is a recessive disorder, since almost all affected sib pairs are HLA identical. The hemochromatosis locus is in close proximity to the HLA-A locus (Edwards et al., 1980). Recent family studies indicate that there are two loci involved in 21-OH deficiency, one located between HLA-B and HLA-DR and the other outside HLA-DR (Dupont et al., 1980). It is important to note, however, that this syndrome can also result from deficiency of 11-OH and 17-OH, the loci for which are not linked to HLA (Brautbar et al., 1979; Mantero et al., 1980).

III. MECHANISMS

Apart from lupus-like syndromes (C2 deficiency) and congenital adrenal hyperplasia (21-OH deficiency), the mechanisms behind HLA disease associations are largely unknown. However, several mechanisms (not mutually exclusive) have been proposed to account for these associations.

1. *Molecular mimicry:* Disease-causing pathogens and HLA antigens may cross-react, so that the organism is unable to distinguish between self and nonself antigens and therefore cannot mount an immune response to eliminate the pathogen.

2. *Receptor:* The HLA antigen in question may actually be a receptor for the disease-causing pathogen(s).

3. *Metabolic genes:* Genes with no immunological role but in linkage disequilibrium with various alleles of the HLA loci may be involved. Such genes could intervene in the metabolic processes connected with the polygenic disease (e.g., hemochromatosis and congenital adrenal hyperplasia).

4. *Immune response (Ir) and immune suppression (Is) genes:* Since most of the associations seem primarily to involve HLA-D/DR antigens, Ir and Is genes (Sasazuki et al., 1980) are most favored candidates as being responsible for HLA and disease associations. These genes must be of great biological importance as a defense mechanism against microbial invasions (e.g., see HLA-DR associated disorders in Table 2).

5. *HLA as a marker for abnormal differentiation antigens:* It has been suggested that some human disease states may result from abnormal embryogenesis controlled by alleles within a human equivalent of the mouse T/t complex. Because of linkage disequilibrium between various specificities of the postulated T/t complex and HLS, variations in normal HLA specificity distribution would be observed. Association between HLA-Dw7 and human testicular teratocarcinoma may be an example of this mechanism (DeWolf et al., 1980).

IV. ACKNOWLEDGEMENTS

We thank Mr. Charles L. Smith for editorial assistance and Dr. Jean Michel Goust for valuable comments.

V. REFERENCES

Bodmer, W. F. (1980). The HLA system and disease. *J. Royal Coll. Phys.* London, 14, 43-50.

Bodmer, W. F. and Payne, R. (1966). Theoretical consideration of leucocyte grouping using multispecific sera. *In* "Histocompatibility Testing 1965." (h. Balner, ed.) pp. 141-149. Munksgaard, Copenhagen.

Braun, W. E. (1979). "HLA and Disease: A Comprehensive Review." CRC Press, Inc., Boca Raton, Florida.

Brautbar, C., Rosler, A., Landau, H. et al. (1979). No linkage between HLA and congenital adrenal hyperplasia due to 11-hydroxylase deficiency. *N. Engl. J. Med.*, 300, 205-206.

Clerget-Darpoux, F. and Bonaiti-Pellié, C. (1980). Epistasis effect: an alternative to the hypothesis of linkage disequilibrium in HLA associated diseases. *Ann. Hum. Genet.*, 44, 195-204.

Darlington, C. D. and Mather, K. (1969). The Elements of Genetics. Schoken Books, N.Y.

Dausset, J. (1977). HLA and association with malignancy: A critical view. *In* "HLA and Malignancy" (G. P. Murphy, ed.) pp. 131-144. Alan R. Liss, Inc., N.Y.

Day, N. E. and Simons, M. J. (1976). Disease susceptibility genes - their identification by multiple case family studies. *Tissue Antigens*, 8, 109-119.

DeWolf, W. C., Dupont, B., and Yunis, E. J. (1980). HLA and disease: current concepts. *Hum. Pathol.*, 11, 332-337.

Dupont, B., Pollack, M. S., Levine, L. S., O'Neill, G. J., Hawkins, B. R., and New, M. I. (1980). Congenital adrenal hyperplasia. *In* "Histocompatibility Testing 1980" (P. I. Terasaki, ed.) pp. 693-706. UCLA Tissue Typing Laboratory, Los Angeles, CA.

Edwards, C. Q., Cartwright, G. E., Skolnick, M. H., and Amos, D. B. (1980). Genetic mapping of the hemochromatosis on chromosome 6. *Hum. Immunol.*, 1, 19-22.

Falk, J. and Osoba, D. (1971). HLA antigens and survival in Hodgkin's disease. *Lancet*, 2, 1118.

Falk, J. and Osoba, D. (1977). The HLA system and survival in malignant disease: Hodgkin's disease and carcinoma of the breast. *In* "HLA and Malignancy" (G. P. Murphy, ed.) pp. 205-216. Alan R. Liss, Inc., N.Y.

Grumet, F. C., Coukell, A., Bodmer, J. G., Bodmer, W. F., and McDevitt, H. O. (1971). Histocompatibility (HL-A) antigens associated with systemic lupus erythematosus. *N. Engl. J. Med.*, 285, 193-196.

Hartley, H. O. (1955). Some recent developments in analysis of variance. *Comm. Pure Appl. Math.*, 8, 47-72.

Haseman, J. K. and Elston, R. C. (1970). The estimation of genetic variance from twin data. *Behav. Genet.*, 1, 11-19.

Kaufman, J. F., Anderson, R. L., and Strominger, J. L. (1980). HLA-DR antigens have polymorphic light chains and invariant heavy chains as assessed by lysine-containing tryptic peptide analysis. *J. Exp. Med.*, 152, 37s-53s.

Li, C. C. (1976). "First Course in Population Genetics." Boxwood Press, Pacific Grove, CA.

Mantero, F., Scaroni, C., Pasini, C. V., and Fagiolo, U. (1980). No linkage between HLA and congenital adrenal hyperplasia due to 17-alpha-hydroxylase deficiency. *N. Engl. J. Med.*, 303, 530.

Markert, M. L. and Cresswell, P. (1980). Polymorphism of human B-cell alloantigens: evidence for three loci within the HLA system. *Proc. Natl. Acad. Sco. U.S.A.*, 77, 6101-6104.

Morton, N. E. (1955). Sequential tests for the detection of linkage. *Am. J. Hum. Genet.*, 7, 277-318.

Rieger, R., Michaelis, A., and Green, M. M. (1976). "Glossary of Genetics and Cytogenetics." 4th ed. Springer-Verlag, N.Y.

Robinson, P. J., Graf, L., and Sege, K. (1981). Two allelic forms of mouse beta$_2$-microglobulin. *Proc. Natl. Acad. Sci.*, 78, 1167-1170.

Sasazuki, T., Kaneoka, H., Nishimura, Y., Kaneoka, R., Hayama, M., and Ohkuni, H. (1980). An HLA-linked immune suppression gene in man. *J. Exp. Med.*, 152, 297s-313s.

Suarez, B. K. (1978). The affected sib pair IBD distribution for HLA-linked disease susceptibility genes. *Tissue Antigens*, 12, 87-93.

Svejgaard, A., Morling, N., Platz, P., Ryder, K. P., and Thomsen, M. (1980). HLA and disease. *In* "Immunology 80." (M. Fougereau and J. Dausset, eds.) pp. 530-540. Academic Press, N.Y.

Terasaki, P. I., Park, M. S., Bernoco, D., Opelz, G., and Mickey, M. R. (1980). Overview of the 1980 International Histocompatibility Workshop. *In* "Histocompatability Testing 1980." (P. I. Terasaki, ed.) p. 15. UCLA Tissue Typing Laboratory, Los Angeles, CA.

vanRood, J. J., deVries, R. R. P., and Bradley, B. A. (1981). Genetics and biology of the HLA system. *In* "The Role of Histocompatability Complex in Immunobiology." (M. E. Dorf, ed.) pp. 59-113. Garland STPM Press, N.Y.

Whittingham, S., Mathews, J. D., Schanfield, M. S., Mathews, J. V., Tait, B. D., Morris, P. J., and Mackay, I. R. (1980). Interactive effect of Gm allotypes and HLA-B locus antigens on the human antibody response to a bacterial antigen. *Clin. Exp. Immunol.*, 40, 8-15.

Whittingham, S., Mathews, J. D., Schanfield, M. S., Tait, B. D., and Mackay, I. R. (1981). Interaction of HLA and Gm in autoimmune chronic active hepatitis. *Clin. Exp. Immunol.*, 43, 80-86.

Woolf, B. (1955). On estimating the relation between blood group and disease. *Ann. Hum. Genet.*, 19, 251-253.

GENERAL CONCEPTS REGARDING MEMBRANE AND ORGANELLE BIOSYNTHESIS

David D. Sabatini, Gert Kreibich, Milton Adesnik, and Takashi Morimoto

Department of Cell Biology
New York University Medical Center
550 First Avenue
New York, New York 10016

Recent studies on organellar and membrane biogenesis have provided substantial evidence for the existence of several mechanisms responsible for guiding newly synthesized polypeptides to the appropriate membranes and, whenever necessary, assisting them in traversing the membrane. In an attempt to relate these events to general concepts within the framework of membrane biogenesis, it is useful to divide the proteins intended for membranes and organelles into two large categories.

Polypeptides which belong to the first category are characterized by the fact that upon release from ribosomes they assume the configuration required for their interaction with the membrane. Their incorporation into the membrane is therefore a post-translational event mediated by the specific recognition by membrane sites of characteristic structural features of the newly synthesized polypeptides. Examples of such proteins are most mitochondrial, glyoxysomal, peroxisomal, chloroplast proteins, as well as certain integral membrane proteins exposed on the cytoplasmic side of the ER membrane, such as cytochrome b_5 and its reductase.

The second category of proteins includes those which are synthesized in ribosomes attached to the endoplasmic reticulum membranes and insert into the ER membrane cotranslationally. Appropriate examples are secretory and lysosomal proteins, as well as integral membrane proteins and glycoproteins which exhibit a transmembrane disposition or are exposed on the luminal or extracellular face of membranes. Such proteins are found in the ER (e.g. cytochrome P450, its reductase, epoxide hydratase, and the Ca^{2+} ATPase of the sarcoplasmic reticulum) or in the plasma membrane (e.g., band 3, the anion channel of the red cell and the glycoproteins of viral envelopes).

ROUTES OF ASSEMBLY OF MYELIN PROTEINS[1]

S. Greenfield, *P. E. Braun, **G. Gantt and E. L. Hogan

Department of Neurology, Medical University of South Carolina
171 Ashley Avenue, Charleston, South Carolina 29425
** P. E. Braun, Department of Biochemistry, McGill University
Montreal, Canada; ***G. Gantt and E. L. Hogan, Department of
Neurology , Medical University of South Carolina, 171 Ashley
Avenue, Charleston, South Carolina 29425

[1]This research was sponsored by Grant NS 12044 from the U.S.
Public Health Service.

I. INTRODUCTION

Membrane genesis has become a focus of intense investigation in recent years. Although many of the mysteries of these processes are beginning to be unraveled, relatively little of this information has resulted from studies of myelinogenesis even though the relative simplicity of myelin and its abundance in brain suggest that studies of the biosynthesis of myelin might explain much about membrane generation. The ultimate importance of studying myelinogenesis derives from the fact that several of the major diseases of the nervous system involve myelin breakdown and that the failure to remyelinate fundamentally may involve a problem in membrane genesis.

The aim of our research is to follow the intracellular route of assembly of basic proteins (BP) and proteolipid proteins (PLP) into the myelin membrane. Our prediction, which can be tested, is that proteolipid protein, an integral membrane protein, follows a mechanism of assembly into the oligodendroglial plasma membrane based on the "membrane flow" model which describes the intracellular transport of proteins from their site of synthesis, through various intracellular compartments, to fusion with the plasmalemma (for reviews, see Rothman and Lenard, 1977; Wickner, 1979; Morré et al., 1979). This model is based on compelling data for the intracellular transport of secretory proteins (for review, see Palade, 1975). The basic proteins, on the other hand, are rather small, soluble peripheral membrane proteins, and their route of assembly probably results from synthesis on free polysomes followed by specific associations with the myelin membrane.

II. METHODS

We have used a combination of subcellular fractionation and double label methodology to explore the route of assembly of myelin basic proteins (MBP) and proteolipid protein from their sites of biosynthesis, through various intracellular compartments, to incorporation into the myelin membrane (Braun et al., 1980). While our results are as yet preliminary, we have identified a number of intracellular membrane fractions related to the process of myelin membrane assembly.

Myelin-related fractions have been identified from mouse brain by the following criteria: (1) SDS-PAGE profiles which include proteins that co-migrate with typical myelin proteins; (2) relative enrichment in DOC-activatable 2', 3' cyclic nucleotide 3' phosphohydrolase (CNP); and (3) relative enrichment of MBPs measured by RIA. Morphological and biochemical characterization of these fractions (Table 1) is being performed by Pedro Pereyra in Dr. Braun's laboratory.

Table 1. Morphological characteristics of subfractions of mouse brain.

FRACTION	MORPHOLOGICAL CHARACTERISTICS
P1Aα	Large fragments of compact lamellar myelin
P1Bα	Large fragments of non-compact myelin and single membranes; some axons
P2Aα	Fragments of compact myelin, myelinated axons and smaller vesicular membranes
P2Aγ	Golgi elements, flattened membranes, heterogeneous vesicular membranes
P2Bα	Large fragments of myelin with few lamelle, no axons
P2Bβ	Myelinated axons and swollen myelin with frequent loops of entrapped cytoplasm resembling paranodes and mesaxonal regions
P2Bγ	Golgi elements, flattened membranes
P3Aα-	Small, smooth vesicles only
P_3D	Many membrane-bound ribosomes

The possible existence of a precursor-product relationship between these fractions and lamellar myelin has been examined by sequential incorporation of amino acid precursors into proteins common to these fractions. A time-staggered double isotope methodology similar to that described by Benjamins and Morell (1978) was used because of its advantages in avoiding problems associated with determination of protein-specific activities.

Thirty 16-day-old mice were injected intracranially with ^{14}C-leucine (250 μCi total). After 60 min, the same mice received ^{3}H-leucine (2500 μCi total) by the same route. Fifteen min later, the mice were decapitated and the brains were immediately homogenized in 0.25 M sucrose, containing 1 mM unlabeled leucine. The homogenate was then fractionated and proteins of the relevant fractions were separated by SDS-PAGE. The stained bands of myelin proteins were excised from the slab gel, digested overnight in Protosol at 50°C, and radioactivity counted in a liquid scintillation spectrometer. Preliminary experiments with varying time intervals established that during the first 60 min, the first isotope (^{14}C) travels from the site of protein synthesis, through the various intracellular compartments (myelin-related fractions) and finally to the site of incorporation into myelin. The second isotope (^{3}H) traverses the same route, but in 15 min it does not label all of the fractions to the same extent as the first isotope. Thus, a given protein will have the highest ratio of ^{3}H to ^{14}C near its site of synthesis, with progressively lower ratios appearing in fractions (intracellular compartments) which are temporarily further removed in the myelination process.

III. RESULTS

After synthesis in microsomes, the first membrane fraction in which proteolipid protein (PLP) appears is P2Bγ (Table 2), a fraction which is enriched in elements of the Golgi complex. The next highest ratio is found in fraction P2Bβ, which also has the largest total amount of labeled PLP. In addition to the presence of morphologically identifiable myelin fragment, this fraction also contains what appear to be paranodal loops.

Table 2. Order of assembly of proteolipid protein.

	$^{3}H/^{14}C$	Total ^{3}H (dpm)
P2Bγ	20	3700
Golgi?		
↓		
P2Bβ	10	5100
Paranodes?		
↓		
P2Bα	3	1100
"New" myelin		
↓		
PIAα	2	110
"Old" myelin		

The P2Bα is the next fraction by order of decreasing $^{3}H/^{14}C$ ratio in PLP. The fraction has morphological characteristics of recently formed ("new") myelin. Within the time frame of our experiment, this is very nearly the end product, although some PLP is observed in P1Aα (compact, mature, or "old" myelin, with a drop in total ^3H dpm and a small decline in ratio).

When ^3H and ^{14}C in the 18.5K basic protein of myelin was determined (Table 3), we observed that both isotopes were incorporated into this protein more rapidly than was observed for PLP. A similar lag in incorporation of PLP into myelin relative to basic protein was first observed by Benjamins and Morell (1978).

Table 3. Order of assembly of 18.5K MBP.

	$^3H/^{14}C$	Total ^3H (dpm)
P3Aα-	9.4	800
Small vesicles		
↓		
P2Bβ + PIBα	8.6	2500
Paranodes + "new" myelin		
↓		
P2Bα	7.6	2800
"New" myelin		
↓		
PIA	7.2	300
"Old" myelin		

It was found that the ratio of ^3H to ^{14}C decreased from 9.4 in the P3Aα-to 7.2 in the myelin end product (Table 3). The high ratio for basic protein in the microsomal fraction (P3Aα ') and the knowledge that basic protein accumulation in this fraction parallels the onset of myelinogenesis and that this fraction is impermeant to ^{14}C sucrose during homogenization (Braun et al., 1980) suggests that this fraction is an intermediate or precursor during myelinogenesis. Two other fractions, P2Bβ and P1Bα, contain membrane compartments through which this basic protein must pass preceding its incorporation into "new" and "old" myelin. The ratios in these two fractions are too close to the same value to permit us to assign a temporal sequence. However, other experiments using a shorter time interval for labeling with the second isotope (^3H) may allow these and earlier events to be distinguished with greater resolution. It is clear from these results that basic protein and proteolipid protein are incorporated into myelin at significantly different rates and by separate routes.

The small basic protein of the mouse (14K MBP; Table 4) also appears first in the microsomal fraction P3Aα', judging by the

Table 4. Order of assembly of 14K MBP.

	$^3H/^{14}C$	Total ^3H (dpm)
P3Aα- + P2Bβ	8.8	1,300
Small vesicles + paranodes?		
↓		
P2Bα + PIBα	7.7	10,600
"New" myelin		
↓		
PIAα	6.9	600
"Old" myelin		

highest ratio (8.8), but another fraction, P2Bβ, also contains this protein with the same labeling ratio. The bulk of the labeled protein then appears in the newly synthesized myelin fraction, with very little mature myelin (P1Aα) becoming labeled in this short time. Again, as with the 18.5K MBP, the second labeling interval (15 min) was not short enough to provide a steep gradient of ratios, so that clear resolution of the early events is not possible.

It should be noted that in the experiments described above the designation of a protein band on SDS slab gels as basic or proteolipid protein is based on a co-migration of these proteins with myelin protein standards. The problem with this is that ambiguous data are obtained if co-migration of one or more proteins with basic or proteolipid protein occurs. In order to increase the degree of confidence in assignment of a protein band as proteolipid protein, we subjected an aliquot of each subfraction to initial extraction with chloroform-methanol (2:1) and precipitated the proteolipid protein with diethyl ether (Greenfield et al., 1979). The isolated proteins were subjected to electrophoresis on 15-25% acrylamide slab gels and bands co-migrating with myelin proteolipid protein were excised and counted for ^3H and ^{14}C activity. The ratios obtained for proteolipid protein by this procedure were compared with the ratios for the unextracted subfractions (Table 5) in the same experiment.

The only fraction that gave comparable $^3H/^{14}C$ ratios was the membrane-bound ribosomal fraction (P3D). The other fractions along the flow route for PLP gave consistently lower $^3H/^{14}C$ ratios for the extracted proteolipid protein. This indicates contamination

Table 5. Order of assembly of proteolipid protein.

	$^{3}H/^{14}C$	$^{3}H/^{14}C$ (2:1 C:M extracted)
P3D	20.9	21.5
Membrane-bound ribosomes ↓		
P2Bγ	17.4	10.3?
Golgi? ↓		
P2Bβ	4.0	1.9
Paranodes ↓		
P2Bα	0.91	0.38
"New" myelin		

with a co-migrating protein, although it should be noted that the order of flow from P3D to P2Bα was not changed. The uncertainty in reporting a value for PLP extracted from P2B γ lies in very low count recovery and difficulty in determining its migration. Clearly, this experiment is preliminary and bears repeating along with additional experiments which add acid-extract for the basic proteins.

These results obtained by chemical extraction and electrophoresis are now being tested by rigorous identification of basic proteins and PLP in each subfraction with a combination of electrophoretic-transfer of proteins from slab gels to nitrocellulose sheets followed by immuno-staining of the immobilized proteins in the solid-phase support (Towbin et al., 1980).

In summary, we have preliminary evidence that myelin assembly is a multi-step process involving at least two different routes for basic and proteolipid protein, occurring at different rates. The demonstration of discrete intracellular compartments in the assembly of the myelin membrane indicates that completing the detailed picture of individual events in this process, such as intracellular transport, soring and membrane fusion is now feasible.

IV. ACKNOWLEDGEMENT

We would like to thank Mrs. Clare Swent for assistance with this manuscript.

V. REFERENCES

Benjamins, J. A. and Morell, P. (1978). Proteins of myelin and their metabolism. *Neurochem. Res.*, 3, 137-174.

Braun, P. E., Pereyra, P. M., and Greenfield, S. (1980). Mechanisms of assembly of myelin in mice: A new approach to the problem. In INSERM Symposium 14, "Neurological Mutations Affecting Myelination." (N. Baumann, ed.) pp. 413-421. Elsevier North-Holland, Amsterdam.

Greenfield, S., Williams, N. I., White, M., Brostoff, S. W., and Hogan, E. L. (1979). Proteolipid protein: Synthesis and assembly into quaking mouse meylin. *J. Neurochem.*, 32, 1647-1651.

Morré, P. J., Kartenbeck, J., and Franke, W. W. (1979). Membrane flow and interconversions among endomembranes. *Biochim. Biophys. Acta*, 559, 71-152.

Palade, G. E. (1975). Intracellular aspects of the process of protein synthesis. *Science*, 189, 347-358.

Rothman, J. E. and Lenard, J. (1977). Membrane asymmetry. *Science*, 195, 743-753.

Towbin, H., Staehlin, T., and Gordon, J. (1979). Electrophoretic transfer of proteins from polyacrylamide gels to nitrocellulose sheets: Procedure and some applications. *Proc. Natl. Acad. Sci. USA*, 76, 4350-4354.

Wickner, W. (1979). The assembly of proteins into biological member trigger hypothesis. *Rev. Biochem.*, 48, 23-45.

MEMBRANE LECTINS OF LYMPHOCYTES[1]

Janet M. Decker

Department of Biochemistry
Medical University of South Carolina
171 Ashley Avenue
Charleston, South Carolina 29425

I. INTRODUCTION

Lectin molecules with binding specificity for several different saccharides have been detected on the membranes of a variety of mammalian cells, including lymphocytes (Stockert et al., 1974; Nowark et al., 1977; Kawasaki and Ashwell, 1977; Kolb and and Kold-Bachofen, 1978; Kawasaki et al., 1978; Stahl et al., 1978; Kieda et al., 1978; Beyer et al., 1979; Briles et al., 1979; Sarkar et al., 1979; Grabel et al., 1979; Baenziger and Maynard, 1979; Steer and Clarenberg, 1979; Kolb et al., 1979; Kieda et al., 1979; Pitts and Yang, 1980; Nagamura and kolb, 1980; Decker, 1980).

[1]Supported in part by the Medical University of South Carolina Biomedical Research Appropriation, 1980-81.

Membrane lectins have been postulated to play a significant role in cellular recognition since (1) they are accessible on the cell surface; (2) many membrane proteins are glycosylated and will react with lectins on other cells (Bretscher, 1973; Nicolson and Singer, 1974), and (3) great oligosaccharide structural diversity is possible due to their branched structure and the varied linkages that can occur. This communication describes molecules isolated from lymphocytes by their affinity binding to the fetal α globulin, fetuin, which is highly glycosylated; the binding is shown to be specific for the oligosaccharide portion of the fetuin molecule.

II. MATERIALS AND METHODS

Thymocytes from Balb/c mice and splenocytes from outbred nu/nu athymic mice, 6-10 weeks of age, were freed of erythrocytes and dead cells and washed in modified Hank's Balanced Salt Solution (MHBSS), pH 7.3, in which 10 mM HEPES (Schwarz/Mann) was substituted for bicarbonate. Adherant peritoneal macrophages were recovered by peritoneal washing 4 days after intraperitoneal injection of 2 ml of thioglycollate medium and enriched by binding to plastic culture dishes at 37°C. A murine B cell line (WEHI 279), grown in continuous culture, was washed three times with MHBSS to free it of fetal calf serum before use. Cells were extracted in ice cold Triton X-100, 1 ml/10^7 cells, without further treatment or treated as described below.

Aliquots of 10^7 cells were iodinated by the lactoperoxidase-catalysed iodination method as described by Marchalonis et al. (1971). After being labeled for 15 min at 30°C, cells were washed twice in phosphate-buffered saline, pH 7.2, to remove unbound ^{125}I and extracted as described below.

Cell concentration was adjusted to 5 x 10^6 cells/ml in Earle's balanced salt solution minus methionine (EBSS minus met), pH 7.3, containing the following supplements per 100 ml medium: 4.9 ml 7.5% $NaHCO_3$; 1 ml each GIBCO 100X MEM vitamins, nonessential amino acids, pyruvate and penicillin-streptomycin; 1 ml each of the GIBCO Selectamine amino acids arginine, cystine, glutamine, histidine, isoleucine, leucine, lysine, phenylalanine, threonine, tryptophan, tyrosine, and valine; 10 mM HEPES: 5 x 10^{-5} M 2-mercaptoethanol; and 1% heat-inactivated fetal calf serum (FCS). The cells were cultured for 16-20 hr at 37°C in a 12% CO_2 humidified atmosphere with 100 µCi ^{75}selenomethionine (^{75}Se-met, 1 mCi/ml, 11.1 Ci/µg, Amersham, Bucks, U.K.), washed free of culture medium, and extracted as described below.

Cell concentration was adjusted to 5 x 10^6 cells/ml in RPMI 1640 (GIBCO) supplemented with 5% FCS and with 2-mercaptoethanol, penicillin-streptomycin, HEPES, and glutamine as described above for the EBSS minus met. Thymocytes were stimulated with 1-2 μg/ml Con A (Calbiochem, La Jolla, Cal.) and spenocytes were stimulated with 50 μg/ml *E. coli* lipopolysaccharide (Sigma, St. Louis, Mo.) for 16-24 hr, freed of culture medium, and resuspended in EBSS minus met for ^{75}Se-met labeling as described in the previous section.

Labeled cells were suspended at 10^7 cells/ml in TBS-Ca (150 mM NaCl, 50 mM Tris, 10 mM $CaCl_2$ and 0.005% merthiolate, pH 8) containing 0.05% Triton X-100 (Calbiochem, La Jolla, Cal.) for 30 min on ice. The extract was then centrifuged to remove unextracted material; occasionally dialysis overnight in the cold against two changes of TBS-Ca was carried out. Aliquots of extract were incubated with 100 μl packed fetuin-Sepharose (conjugated at a ratio of 10 mg fetuin/ml CNBr-activated Sepharose 4B) for 30 min at room temperature; the adsorbent was washed three times with 1 ml of TBS-Ca and the bound material was eluted with reducing sample buffer [62 mM Tris-HCl, pH 6.8, containing 15% glycerol, 3% sodium dodecyl sulfate (SDS), and 5% 2-mercaptoethanol] at 100°C for 3-5 min and analyzed by polyacrylamide gel electrophoresis in SDS-containing buffers (SDS PAGE) according to the method of Laemmli (1970).

Sepharose 4B CL and CNBr-activated Sepharose 4B were purchased from Pharmacia, Uppsala, Sweden; fetuin and bovine submaxillary mucin from Sigma Chemical Co., St. Louis, Mo.; and BSA from Calbiochem, La Jolla, Cal.

III. RESULTS

Fetuin is an α-globulin found in fetal calf serum; it accounts for 40-50% of total fetal serum protein (10-22 mg/ml) depending on gestational age (Graham, 1972). The major fetuin oligosaccharide is a complex branched sugar with terminal sialic acid residues linked via galactose and N-acetyl-glucosamine to a mannose and N-acetyl-glucosamine backbone and coupled to the protein via asparagine residues (Baenziger and Fiete, 1979; Nilsson et al., 1979). Molecules that bind to the fetuin oligosaccharide have been detected in extracts of murine thymocytes and splenocytes using fetuin-Sepharose adsorbents as described under Materials and Methods. Figure 1 illustrates the SDS-PAGE analysis of surface iodinated molecules from nu/nu mouse splenocytes (B lymphocytes) which bind to fetuin sepharose. The major peak of binding material is found migrating slightly ahead of an external ovalbumin standard, with another band of higher mobility and minor bands migrating just ahead of immunoglobulin light chain and just behind ovalbumin. Minimal binding is seen with unconjugated Sepharose 4BCL. The presence of 10 mg/ml free fetuin (a tenfold

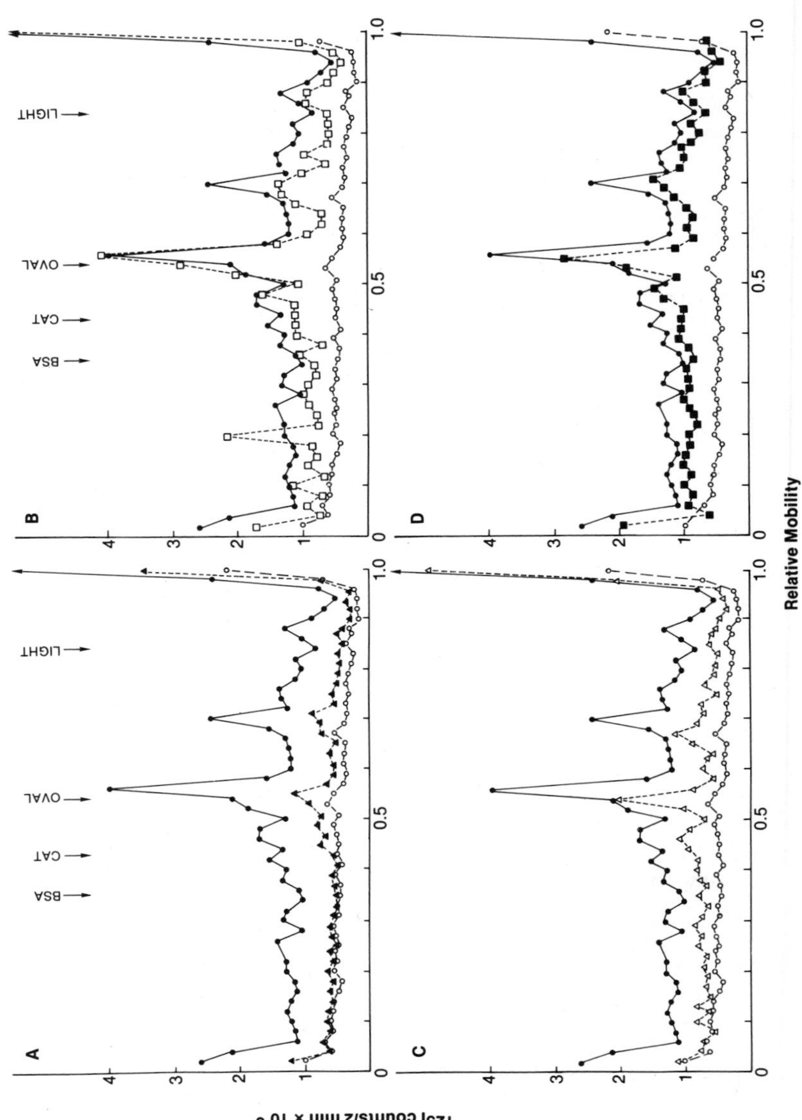
A
B
C
D
BSA
CAT
OVAL
LIGHT
125I counts/2 min × 10-3
Relative Mobility
0
0.5
1.0
1
2
3
4

Fig. 1. SDS-PAGE analysis of fetuin-sepharose binding proteins from murine splenocytes. Spleen cells from nu/nu athymic mice were radioiodinated, extracted with Triton X-100, and bound to fetuin-sepharose in the presence or absence of added protein. The bound material was eluted in reducing sample buffer and analyzed on rod gels (10% acrylamide, 0.5% bisacrylamide). which were then sliced into 2mm segments and counted for ^{125}I. External molecular weight standards were bovine serum albumin (BSA, 68,000 daltons); catalase (CAT, 60,000 daltons); ovalbumin (OVAL, 43,000 daltons); and immunoglobulin light chain (LIGHT, 23,000 daltons). A-D. ●——● bound to fetuin sepharose in the absence of competing protein. ○——○ bound to unconjugated Sepharose 4B CL. Other curves show binding to fetuin-sepharose in the presence of A. ▲- - -▲ 10 mg/ml free fetuin. B. □- - -□ 10 mg/ml BSA. C. △- - -△ mg/ml bovine submaxillary mucin. D. □- - -□ 10 mg/ml human IgG.

excess over the bound fetuin present on the absorbant) blocked extract binding to fetuin sepharose. Bovine submaxillary mucin, which contains about 60% by weight O-glycosidically linked oligosaccharide with a high content of sialic acid, was also a good inhibitor of binding to fetuin sepharose. Nonglycosylated bovine serum albumin (BSA), and human immunoglobulin G (IgG) with high mannose containing oligosaccharides, were poor inhibitors of fetuin binding by lymphocyte extracts. Similar patterns of binding and inhibition were seen with extracts of Balb/c thymocytes (Decker, 1980). Glycoproteins with complex oligosaccharides similar in structure to the fetuin oligosaccharides, such as bovine submaxillary mucin, throglobulin, and human α-1 acid glycoprotein (orosomucoid), as well as gangliosides from bovine brain, were better inhibitors than glycoproteins with dissimilar oligosaccharides such as ovalbumin and human IgG or nonglycosylated proteins such as *Staphylococcus* Protein A. Monosaccharides were in general not good inhibitors of the affinity binding; the only significant inhibition obtained was with N-acetyl-neuraminic acid (sialic acid) which in one experiment was as high as 85% with 100 mM sugar.

Several other experiments were performed to investigate the specificity of the binding to fetuin sepharose. Fetuin digested with pronase using the conditions of Spiro and Bohoyroo (1974) in which the protein was digested while the oligosaccharides remained intact, was a good blocker of fetuin-sepharose binding. The glycopeptides of fetuin, approximate size 4000 daltons, were purified from the pronase digest by Sephadex G50 chromatography in Tris acetate buffer, pH 7.8, and these glycopeptides were conjugated to CNBr-activated sepharose. Figure 2 illustrates that the same binding molecules were observed in an extract of the murine B cell line WEHI 279 with glycopeptide-sepharose as with fetuin-sepharose. As with the iodinated material, the major band of eluted material migrates just ahead of an external ovalbumin standard, while higher and lower mobility bands are also present. Contamination of the eluate with fetuin leaching from the adsorbent (seen in gels 3 and 4) is not a problem with the glycopeptide-sepharose adsorbant. Further work is underway to achieve a single band by SDS-PAGE following affinity chromatography; however, the lectin binding by the glycopeptide sepharose is additional evidence that the binding specificity of the lymphocyte material is directed aginst oligosaccharide and not peptide determinants.

To investigate the fetuin-binding material from activated lymphocytes it was necessary to label the cells biosynthetically, since the mitogen-stimulated cells were too aggregated for effective iodination. Figure 3 shows that the same major binding pattern was achieved using material from ^{75}selenomethionine-

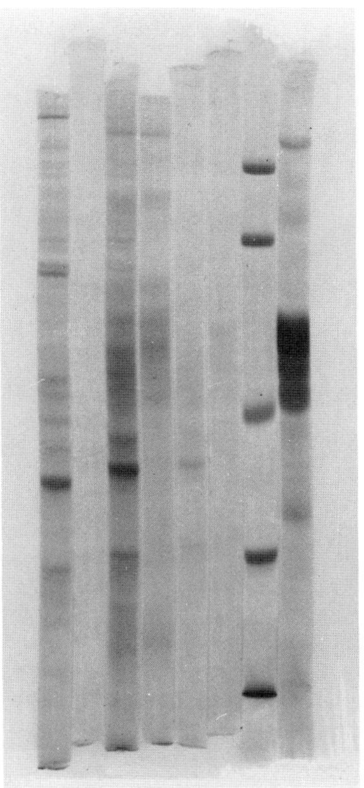

Fig. 2. Lymphocyte extract binding to fetuin glycopeptides conjugated to Sepharose 4B. Aliquots of Triton X-100 extract from WEHI 279 were bound to glycopeptide-sepharose, fetuin-sepharose, or unconjugated Sepharose 4B CL; bound material was eluted with reducing sample buffer, analyzed by SDS-PAGE, and stained with Comassie Blue. Control gels of material eluted from sepharose which had been incubated only with buffer were also included to check for leaching of protein from the affinity absorbants. From left to right, the gels represent material eluted from (1) glycopeptide sepharose incubated with WEHI 279 extract, (2) glycopeotide-sepharose incubated with buffer, (3) fetuin-sepharose incubated with WEHI 279 extract, (4) fetuin-sepharose incubated with buffer, (5) sepharose 4B CL incubated with WEHI 279 extract, (6) sepharose 4B CL incubated with buffer, (7) external molecular weight standards (from top) phosphorylase b, 94,000 daltons; BSA, 68,000 daltons; ovalbumin, 43,000 daltons; carbonic anhydrase, 30,000 daltons, and soybean trypsin inhibitor, 20,100 daltons; (8) Sigma Type IV fetuin, 10 μl of a 10 mg/ml solution.

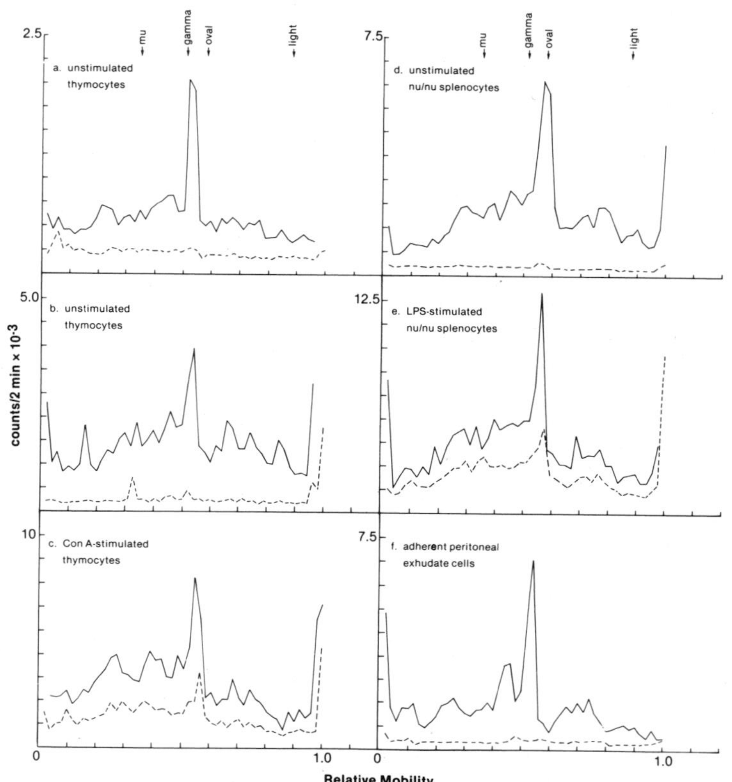

Fig. 3. Binding of biosynthetically-labeled lymphocyte proteins to fetuin-sepharose and horse erythrocytes. Balb/c thymocytes and adherant peritoneal exudate cells, and spenocytes from nu/nu athymic mice, were labeled overnight with ^{75}selenomethionine, which replaces S-methionine in proteins. Following Triton X-100 extraction, the labeled material was incubated with 100 μl packed glutaraldehyde-fixed horse erythrocytes. Bound material was eluted with sample buffer and analyzed by SDS-PAGE. ________ Binding to fetuin-sepharose. ------ Binding to horse erythrocytes. Molecular weights of the external standards are immunoglobulin (Ig) μ chain, 70,000 daltons; Ig γ chain, 50,000 daltons; ovalbumin, 43,000 daltons; and Ig light chain, 23,000 daltons. a, c-f. ^{75}Selenomethionine-labeled cells. b. ^{125}I-labeled thymocytes prepared in the same experiment for comparison. a,b. Unstimulated thymocytes. c. Con A-stimulated thymocytes. d. Unstimulated spenocytes. e. LPS-stimulated splenocytes. Adherant peritoneal exudate cells.

labeled cells as with lactoperoxidase iodinated cells (b). Unstimulated cells yielded very little material which was bound to glutaraldehyde-fixed horse erythrocytes; this material was increased when the cells were stimulated with mitogen before labeling. These results suggest that the fetuin-binding material is synthesized by the lymphocytes rather than being cytophilically adsorbed to the cell surfaces; they also argue for the protein nature of the fetuin-binding material, since it can be labeled with methionine.

IV. DISCUSSION

Affinity purification of Triton X-100 extracts from lymphocytes yields saccharide-binding material of heterogeneous relative mobilities on SDS-PAGE. Several reasons for the presence of multiple bands might exist: (1) There are several lectin molecules on lymphocyte membranes, either with diverse or identical binding specificities. (2) Only one lectin is present on the lymphocyte, and other proteins associated with it in the membrane remain associated in the Triton X-100 extract and are carried along with the true binding material. (3) Proteolytic breakdown of the lectin during the experiment results in multiple bands. The fourth possibility, that the multiple bands represent multiple lectins present on the several cell types present in the thymocyte and splenocyte preparations, seems less likely in view of the multiple bands present in the WEHI 279 material, which is prepared from a monoclonal cell line. Experiments are now underway which are designed to separate the various molecules from each other and test their individual binding to fetuin-sepharose.

Although monosaccharides are not good inhibitors of binding to fetuin-sepharose, the inhibition and binding data with fetuin glycopeptides and the inhibition data with glycoproteins and glycolipids with similar oligosaccharides offers convincing evidence that the binding is directed towards a saccharide determinant on the fetuin. The poor ability of monosaccharides to block binding probably indicates that the lectin recognizes a determinant formed of several saccharides; possibly the linkage between individual sugars is critical for recognition. Plant lectins once thought to recognize monosaccharides have later been shown to recognize larger determinants.

Fetuin was chosen for this work because its complex oligosaccharides offered the best chance to detect binding directed at many diverse saccharide determinants, yet the structure was known so that the lymphocyte lectin binding specificity could be defined. Chemical and enzymatic techniques are being used to modify the fetuin oligosaccharide until the part of the structure bound by the lymphocyte lectin(s) can be identified. Binding specificities of lectins from T and B cells will be compared, and binding of purified lectin to lymphocytes will be done to investigate the possibility that these molecules are involved in the recognition or communication between lymphoid cells. Fetuin-binding molecules have also been found on non-lymphoid cells (Decker, 1980); and they may prove to be recognition molecules present on many different cell types.

V. REFERENCES

Baenziger, J. U. and Fiete, D. (1979). Structure of the complex oligosaccharides of fetuin. *J. Biol. Chem.*, 254, 789-795.

Baenziger, J. U. and Maynard, Y. (1980). Human hepatic lectin. Physiochemical properties and specificity. *J. Biol. Chem.*, 255, 4607-4613.

Beyer, E. C., Tokuyasu, K. T., and Barondes, S. H. (1979). Localization of an endogenous lectin in chick liver, intestine and pancreas. *J. Cell Biol.*, 82, 565-571.

Bretscher, M. S. (1973). Membrane structure: some general principles. *Science*, 181, 622-629.

Briles, E. B., Gregory, W., Fletcher, P., and Kornfeld, S. (1979). Vertebrate lectins. Comparison of properties of β-galactoside-binding lectins from tissues of calf and chicken. *J. Cell Biol.*, 81, 528-537.

Decker, J. M. (1980). Lectin-like molecules on murine thymocytes and splenocytes I. Molecules specific for the oligosaccharide moiety of the fetal α-globin fetuin. *Mol. Immunol.*, 17, 803-808.

Grabel, L. B., Rosen, S. D., and Martin, G. R. (1979). Teratocarcinoma stem cells have a cell surface carbohydrate-binding component implicated in cell-cell adhesion. *Cell*, 17, 477-484.

Graham, E. R. B. (1972). Fetuin. In "Glycoproteins" (Edited by Gottschalk, A.) pp. 717-731. Elsevier, Amsterdam.

Kawasaki, T. and Ashwell, G. (1977). Isolation and characterization of an avian hepatic binding protein specific for N-acetylglucosamine-terminated glycoproteins. *J. Biol. Chem.*, 252, 6536-6543.

Kawasaki, T., Etoh, R., and Yamashina, I. (1978). Isolation and characterization of mannan-binding protein from rabbit liver. *Biochem. Biophys. Res. Commun.*, 81, 1018-1024.

Kieda, C. M. T., Bowles, D. J., Ravid, A., and Sharon, N. (1978). Lectins in lymphocyte membranes. *FEBS Lett.*, 94, 391-396.

Kieda, C., Roche, A.-C., Delmotte, F., and Monsingny, M. (1979). Lymphocyte membrane lectins. Direct visualization by the use of fluorescein-glycosylated cytochemical markers. *FEBS Lett.*, 99, 329-332.

Kolb, H. and Kolb-Bachofen, V. (1978). A lectin-like receptor on mammalian macrophages. *Biochem. Biophys. Res. Commun.*, 85, 678-683.

Kolb, H., Kolb-Bachofen, V., and Schlepper-Schäfer, J. (1979). Cell contacts mediated by D-galactose-specific lectins on liver cells. *Biol. Cellulaire*, 36, 301-308.

Laemmli, U. K. (1970). Cleavage of structural proteins during the assembly of the head of bacteriophage T4. *Nature, Lond.*, 227, 680-685.

Marchalonis, J. J., Cone, R. E., and Santer, V. (1971). Enzymic bioiodination: A probel for accessible surface proteins of normal and neoplastic lymphocytes. *Biochem. J.*, 124, 921-927.

Nagamura, Y. and Kolb, H. (1980). Presence of a lectin-like receptor for D-galactose on rat peritoneal macrophages. *FEBS Lett.*, 115, 59-62.

Nicolson, G. L. and Singer, S. J. (1974). The distribution and asymmetry of mammalian cell surface saccharides utilizing ferritin-conjugated plant agglutinins as specific saccharide stains. *J. Cell. Biol.*, 60, 236-248.

Nilsson, B., Norden, N. E., and Svenson, S. (1979). Structural studies on the carbohydrate portion of fetuin. *J. Biol. Chem.*, 254, 4545-4553.

Nowak, T. P., Kobiler, D., Roel, L. E., and Barondes, S. H. (1977). Developmentally regulated lectin from embryonic chick pectoral muscle. *J. Biol. Chem.*, 252, 6026-6030.

Pitts, M. J. and Yang, D. C. H. (1980). Isolation of a developmentally regulated lectin from chick embryo. *Biochem. Biophys. Res. Comm.*, 95, 750-757.

Sarkar, M., Lio, J., Kabat, E. A., Tanabe, T., and Ashwell, G. (1979). The binding site of rabbit hepatic lectin. *J. Biol. Chem.*, 254, 3170-3174.

Sharon, N. and Lis, H. (1972). Lectins: cell-agglutinating and sugar-specific proteins. *Science*, 177, 949-959.

Spiro, R. G. and Bhoyroo, V. D. (1974). Structure of the O-glycosidically linked carbohydrate units of fetuin. *J. Biol. Chem.*, 249, 5701-5717.

Stahl, P. D., Rodman, J. S., Miller, M. J., and Schlesinger, P. H. (1978). Evidence for receptor mediated binding of glycoproteins, glycoconjugates, and lysosomal glycosidases by aveolar macrophages. *Proc. Natl. Acad. Sci. USA*, 75, 1399-1403.

Steer, C. J. and Clarenburg, R. (1979). Unique distribution of glycoprotein receptors on parenchymal and sinusoidel cells of rat liver. *J. Biol. Chem.*, 254, 4457-4461.

Stockert, R. J., Morell, A. G., and Scheinberg, I. H. (1974). Mammalian hepatic lectin. *Science*, 186, 365-366.

SODIUM-DEPENDENT CALCIUM FLUXES PROMOTED BY SARCOLEMMA VESICLES FROM CARDIAC TISSUE[1]

George E. Lindenmayer, Dieter K. Bartschat, and Robin T. Hungerford

Departments of Pharmacology and Medicine
Medical University of South Carolina
171 Ashley Avenue
Charleston, South Carolina 29425

[1]This work was supported, in part, by Grants GM20387 and HL23802 from the United States Public Health Service.

[2]Predoctoral Trainee supported by Training Grant GM07409 from the United States Public Health Service.

I. INTRODUCTION

Excitation-contraction coupling of the myocardial cell is believed to involve (1) a signal-usually electrical in nature, (2) the movement of calcium across the sarcolemma into the cell, and (3) a transient increase in cytoplasmic free calcium which diffuses into the sarcomere for contraction (Van Winkle and Schwartz, 1976; Fozzard, 1977; Katz, 1977). During diastole, the myocardial cell is electrically polarized (inside negative) and the sarcolemma separates high (millimolar) free calcium in the extracellular space from low (submicromolar) free calcium in the cytoplasm. The signal for contraction depolarizes the cell which is believed to transiently increase sarcolemma permeability to calcium. Two mechanisms that may be involved are activation of a calcium channel through which calcium flows into the cell down an electrochemical gradient (Reuter, 1979) and a switch in the direction of an electrogenic exchange of sodium for calcium such that sodium efflux is coupled to calcium influx (Mullins, 1977).

The recent availability of highly enriched sarcolemma preparations from cardiac tissue (e.g., Jones et al., 1979; Van Alstyne et al., 1980) has allowed an exploration of these events in a biological system that is less complex than intact cells or myocardial tissue preparations. Evidence has been obtained for the existence in such preparations of a sodium-calcium exchange system (Reeves and Sutko, 1979; Pitts, 1979) that is electrogenic (Miyamoto and Racker, 1980; Philipson and Nishimoto, 1980; Reeves and Sutko, 1980). More recently, Bartschat et al. (1980) reported that depolarization of vesicles in a sarcolemma-enriched preparation caused the rapid uptake of calcium. It seemed possible that this reaction reflected the activation of and subsequent movement of calcium through a calcium channel. The reaction, however, required the presence of extravesicular sodium which raised the alternative possibility that the reaction was due to electrogenic sodium-calcium exchange. The experiments described below were designed to determine which of these two possibilities is correct.

II. METHODS

The isolation procedure of Van Alstyne et al. (1980) was employed to obtain osmotically-active vesicular preparations from canine left ventricle. These preparations were enriched to 27 to 40+-fold with putative sarcolemma markers.

A three stage procedure, previously reported (Bartschat et al., 1980) and shown in Figure 1, was used to determine the effects of membrane depolarization on calcium uptake by the vesicles. Briefly, the vesicles were loaded overnight in a medium containing

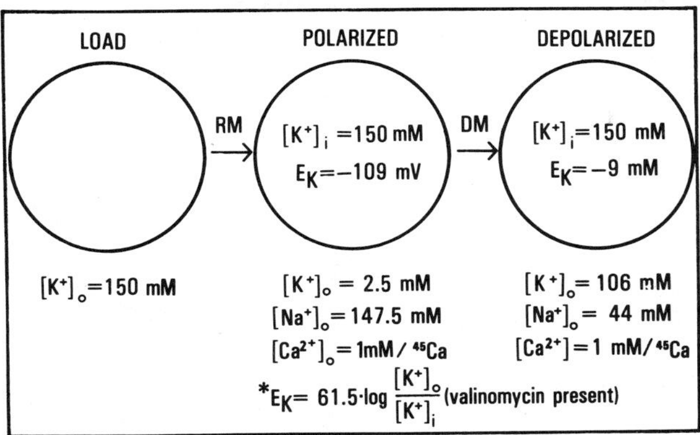

Figure 1. Three stage procedure used to determine the effects of membrane depolarization on calcium uptake by vesicles in a sarcolemma enriched preparation. RM = reaction medium; DM = dilution medium; E_K = potassium Nernst potential which is given by the equation; circles depict the vesicles. Chloride salts were used throughout and all media contained 10 mM Tris-Cl, pH 7.4 (37°C).

high potassium. Next, the suspension was prewarmed to 37°C and added to a reaction medium which contained low potassium in the presence of valinomycin to allow the vesicles to become electrically polarized through the development of a potassium diffusion potential (inside negative). Calcium and ^{45}Ca were also present to allow an assessment of calcium uptake by the vesicles during the polarized state. Finally, the suspension was diluted into a medium containing high potassium. Thus, the diffusion potential for potassium was decreased and the vesicles were correspondingly depolarized. This medium contained the same concentration of calcium and the same specific activity of ^{45}Ca that was present at the previous (polarized) stage so that the effects of depolarization on calcium movement could be determined. One other difference, besides in potassium concentration, between the extravesicular media of the second and third stages is that extravesicular sodium was high in the former but low in the latter. This was necessary in order to keep the ionic strength constant between the two stages. The reactions were terminated by addition of an ice-cold "stopping solution" which contained 150 mM KCl, 1 mM $CaCl_2$, 0.1 mM $LaCl_3$, and 10 mM Tris-Cl, pH 7.4. ^{45}Ca associated with the vesicles was separated from extravesicular ^{45}Ca by filtration and washing of the filter and all data were corrected for ^{45}Ca associated with the filter *per se* (Bartschat and Lindenmayer, 1980). Variants of this protocol and details for each experiment are presented in the figure legends.

III. RESULTS AND DISCUSSION

If one assumes that the vesicular membrane potential is equal to the calculated E_K, depolarization from -109 to -9 mV (Fig. 1) leads to an abrupt rise in vesicular calcium (Fig. 2). As stated above, the two alternative mechanisms most likely to be reflected by such a reaction are a voltage-sensitive calcium channel or an electrogenic sodium-calcium exchange. In fact, several features of the results in Figure 2 suggested that the reaction might reflect calcium movement into the vesicles through a calcium channel (Bartschat et al., 1980). First, as shown by the profiles, the reaction seemed to be quite fast and also appeared to terminate quickly. This would be consistent with voltage- and time-dependent effects on the activation and inactivation parameters of a channel. Conversely, the original studies of sodium-calcium exchange in isolated sarcolemma preparations suggested that calcium movement via the exchange is relatively slow (Reeves and Sutko, 1979; Pitts, 1979). Second, the increase in vesicular calcium upon depolarization did not seem to be due to the decrease in extravesicular sodium, i.e., necessary for the increase in potassium by depolarization. As shown in Figure 2, a similar drop in outside sodium, by dilution with choline instead of potassium, did not significantly affect vesicular calcium. Thus, it appeared that the stimulation of calcium uptake, by increasing extravesicular potassium, is due to depolarization of the vesicles and not due or even related to the decrease in extravesicular sodium. Conversely, a drop in extravesicular sodium should markedly stimulate sodium efflux coupled to calcium influx via sodium-calcium exchange because extravesicular sodium is inhibitory for that reaction.

There were, however, three features which suggested that the reaction was dependent on sodium (Bartschat et al., 1980): (1) the reaction does require the presence of extravesicular sodium whereas a current ascribed to a calcium channel in cardiac tissue does not; (2) studies with the voltage-sensitive dye, DiS-C_3-(5)[3],

[3]DiS-C_3-(5) is 3,3'-dipropyl-2, 2'-thiadicarbocyanine.

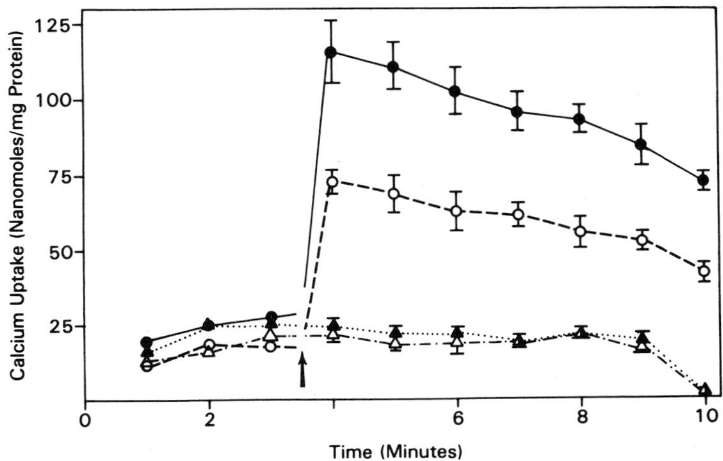

Fig. 2. Effect of transitions in membrane potential on calcium uptake. Vesicles were polarized at zero time and transitions in potential were made at the arrow. Closed and open symbols denote presence and absence of 0.3 μM valinomycin, respectively. At the arrow, extravesicular potassium was either elevated to change E_K from -109 to - 9 mV (depolarization; ●, ○) or decreased to change E_K from -109 to -142 mV (hyperpolarization; ▲, △) but in all cases, extravesicular sodium after the arrow was 44.25 mM (difference in ionic strength made up with choline Cl as needed). All other conditions were as shown in Figure 1. From Bartschat et al. (1980). J. Biol. Chem., 255: 10044.

suggested that E_K gradually dissipates prior to the depolarization step. The likely reason for this time-dependent event is that sodium leaks into the vesicles and displaces (decreases) intravesicular potassium (see the equation in Fig. 1); (3) calcium was found to be concentrated against its chemical gradient upon depolarization although this could have been explained by calcium movement through a channel when membrane potential (E_K), upon depolarization, was kept slightly negative (i.e., at -9 mV) or by an electrogenic sodium-calcium exchange. Overall, therefore, one could not conclude from the initial results that depolarization-induced calcium uptake by the isolated vesicles is due to activation of a calcium channel as opposed to electrogenic sodium-calcium exchange.

Several experiments, reported below, were designed to further test these possibilities. The first focused on whether or not decreases in extravesicular sodium without change in potential could, under certain conditions, induce calcium uptake, i.e., as opposed to the choline dilution experiments in Figure 2. The experimental design was based on the following considerations. Pitts (1979) suggested that the stoichiometry of sodium-calcium exchange in isolated sarcolemma preparations from heart is three sodium per one calcium. From Mullins (1977), the equilibrium equation for such a stoichiometry is:

$$\text{Equation 1.} \quad [Ca^{2+}]i = \frac{[Ca^{2+}] \cdot [Na^{+}]i^{3}}{[Na^{+}]_{o}^{3} \cdot e^{-EmF/RT}}$$

where the subscripts i and o refer to intravesicular and extravesicular concentrations, respectively, F, R, and T have their usual chemical meanings and Em is equated to E_K. If one assumes that $[Ca^{2+}]_o$ = 1 mM, $[Na^+]_o$ = 147.5 mM, $[Na^+]_i$ = 50 mM (due to leak of sodium into the vesicles), and E_K = -98.5 mV (decreased from -109 mV by displacement of potassium to 100 mM by sodium in the intravesicular space), the $[Ca^{2+}]_i$ equals 1 μM prior to depolarization or hyperpolarization (Fig. 1). Now, if $[Na^+]_o$ is reduced to 44.25 mM and E_K is increased to -131 mV (Fig. 2-experiments with choline dilution; assumes intravesicular potassium = 100 mM), $[Ca^+]_i$ rises to 11 μM. Conversely, if $[Na^+]_o$ is reduced to 44.25 mM and E_K is depolarized to 1.5 mV (Fig. 2-experiments with potassium dilution; assumes intravesicular potassium = 100 mM), then $[Ca^{2+}]_i$ rises to 1527 μM which is in the range observed for vesicular calcium after depolarization. These calculations, therefore, suggest that a change from 1 to 11 μM vesicular calcium might have occurred via sodium-calcium exchange in the choline dilution experiments of Figure 2 but such a change might not have been detectable due to the prevailing background of vesicular ^{45}Ca, e.g., due to calcium-calcium exchange (Bartschat and Lindenmayer, 1980). Conversely, increases in vesicular calcium to millimolar levels, predicted for the potassium dilution experiments by Equation 1, were detected. From these and additional calculations, it seemed apparent that changes in vesicular calcium, if induced by alterations in extravesicular sodium, would be more easily observed when the membrane potential remains small or is absent through the course of the experiment.

Accordingly, two experiments were carried out that used vesicles loaded with both sodium and potassium. A much shorter time course was employed to minimize changes in intravesicular sodium and potassium and in E_K due to the sodium leak. In

Experiment B (Fig. 3), the vesicles were polarized to -109 mV (calculated E_K) in the presence of 100 mM extravesicular sodium for 10 sec. Upon depolarization to 0 mV with a drop in outside sodium to 30 mM, vesicular calcium was increased by 4.6-fold to 24.9 ± 2.2 nmoles/mg protein. In a second experiment (Experiment A, Fig. 3), the vesicles were never polarized but a drop in outside sodium from 100 to 30 mM increased vesicular calcium to 23.3 ± 1.4 nmoles/mg protein. Thus, both experiments yielded essentially identical final values for vesicular calcium but in one, there was no transition in membrane potential. This suggests that depolarization-induced calcium uptake, as shown in Figures 2 and 3, does not reflect the activation of a voltage-sensitive calcium channel.

The rationale for the next experiment was based on the following considerations. The calcium channel in cardiac tissue is believed to be activated by depolarization of the cell beyond a threshold of around -40 to -35 mV (Reuter, 1979). Repolarization to potentials more negative than this threshold are associated with conversion of the channel to a state that is nonconducting (closed) but that can be activated (opened) by the next depolarization. If depolarization-induced calcium uptake in Figures 2 and 3 were due to activation of a calcium channel, one can make the following prediction: Depolarization of the vesicles would lead to activation of a calcium channel and calcium uptake by the vesicles but hyperpolarizing the vesicles should have no effect on vesicular calcium. In fact, depolarization with a drop in outside sodium abruptly increased vesicular calcium by 13.8 + 1.7 nmoles/mg protein (C minus B; Fig. 4) but hyperpolarization with an increase in outside sodium abruptly dropped vesicular calcium by 15.8 ± 3.7 nmoles/mg protein (D minus E; Fig. 4). Thus, the depolarization-induced uptake of calcium by the isolated vesicles appears to be entirely reversible. One must conclude from this result that at least most of the uptake, induced by depolarization, is not due to activation of a calcium channel but rather probably reflects an electrogenic exchange of intravesicular sodium for extravesicular calcium.

These results and interpretations suggest that isolated sarcolemma vesicles manifest a quite fast component of electrogenic sodium-calcium exchange which appears to reach equilibrium by at least two seconds (Hungerford and Lindenmayer, 1981). If the reaction, in fact, reaches or even approaches equilbrium within 100 to 500 msec (the time course of the ventricular myocardial cell action potential), it could furnish much of the calcium needed for myocardial contraction and, in turn, could play a key role in effecting relaxation of the cell by subsequently removing cytoplasmic calcium (Equation 1 with physiological concentrations of $[Ca^{2+}]_o$, $[Na^+]_o$, $[Na^+]_i$ and values for the transitions that occur in the ventricular action potential (see Fig. 3 in Fabiato and Fabiato, 1978).

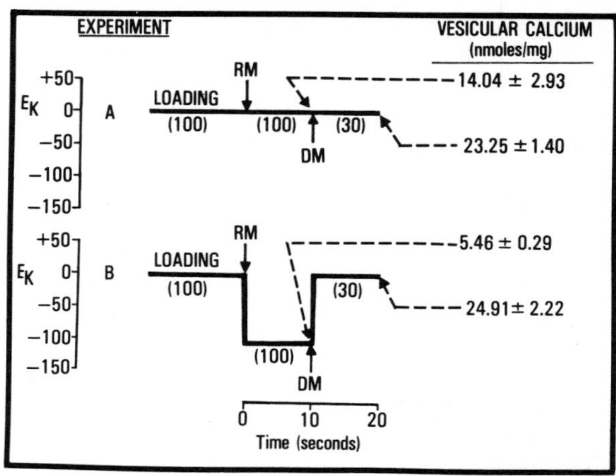

Fig. 3. Effects of change in extravesicular sodium on calcium uptake. Aliquots of sarcolemma enriched preparations were loaded overnight with a medium containing 50 mM KCl, 100 mM NaCl, 50 mM LiCl and 10 mM Tris-Cl, pH 7.4. After prewarming to 37°C, the suspensions were added to a reaction medium (RM) to yield final extravesicular concentrations for Experiment A of 50 mM KCl, 100 mM NaCl, 50 mM LiCl, 0.1 μM valinomycin, 1 mM $CaCl_2$ with ^{45}Ca and 10 mM Tris-Cl, pH 7.4 and for Experiment B of 0.83 mM KCl, 100 mM NaCl, 99.17 mM LiCl, 0.1 μM valinomycin, 1 mM $CaCl_2$ with ^{45}Ca and 10 mM Tris-Cl, pH 7.4. This yielded values for E_K of 0 and -109 mV for Exps. A and B, respectively (E_K depicted by broad, heavy lines; extravesicular sodium concentrations by numbers in parentheses under these lines). Ten seconds later, the reactions were either terminated (upper values of vesicular calcium for each experiment) or diluted to yield final extravesicular concentrations of 50 mM KCl, 30 mM NaCl, 120 mM LiCl, 0.1 μM valinomycin, 1 mM $CaCl_2$ with ^{45}Ca and 10 mM Tris-Cl, pH 7.4. Thus, $E_K = 0$ for both experiments upon dilution. After an additional 10 seconds, the reactions were terminated (lower values of vesicular calcium for each experiment). Each value is the mean ± SE (n = 3).

As Mullins (1977) pointed out, such a system would move calcium in the direction required by the cell for cyclic contractions, viz., inward upon membrane depolarization and outward upon repolarization.

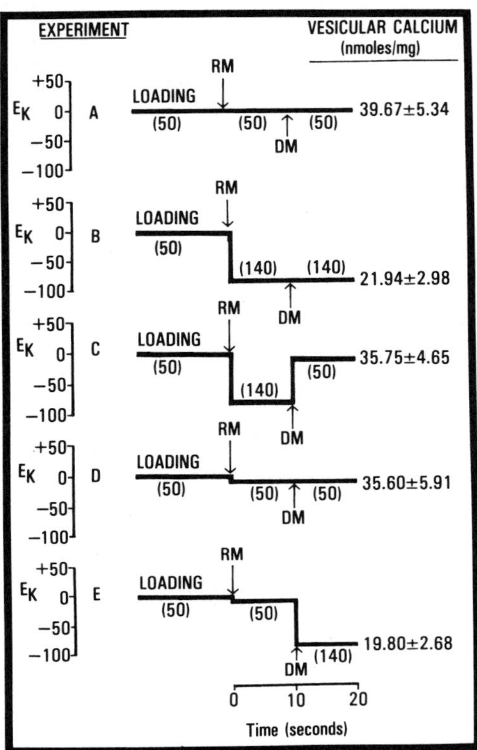

Fig. 4. Effect of sequence of changes in membrane potential and extravesicular sodium concentrations on vesicular calcium. Aliquots of the sarcolemma enriched preparations were loaded overnight with 50 mM KCl, 50 mM NaCl, 100 mM LiCl, 1.0 mM $CaCl_2$ with ^{45}Ca and 10 mM Tris-Cl, pH 7.4. After a brief period of warming to 37°C, the suspensions were added into media containing sufficient potassium (0.1 μM valinomycin present) to yield values of E_K = 0, Experiment A, -9 mV, Experiments D and E, or -80 mV, Experiments B and C (E_K throughout experiments is depicted by heavy broad lines), extravesicular sodium as shown in parentheses under the lines for E_K and 1 mM $CaCl_2$ with ^{45}Ca. After 10 sec, these suspensions were diluted to yield values for E_K and extravesicular sodium that were unchanged, Experiments A, B, and D, to depolarize in Experiment C from -80 to -9 mV with reduction in extravesicular sodium as shown and to polarize in Experiment E from -9 to -80 with change in sodium as shown. After an additional 10 sec, the reactions were terminated. Values for vesicular calcium are means ± SE (N = 4).

IV. ACKNOWLEDGEMENTS

The authors thank Amey Maybank and Eldwin Van Alstyne for isolation of the sarcolemma-enriched preparations.

V. REFERENCES

Bartschat, D. K. and Lindenmayer, G. E. (1980). Calcium movements catalyzed by vesicles in a highly enriched sarcolemma preparation from canine ventricle: Calcium-calcium and sodium-calcium transport. *J. Biol. Chem.*, 255, 9626-9634.

Bartschat, D. K., Cyr, D. L., and Lindenmayer, G. E. (1980). Depolarization-induced calcium uptake by vesicles in a highly enriched sarcolemma preparation from canine ventricle. *J. Biol. Chem.*, 255, 10044-10047.

Fabiato, A. and Fabiato, F. (1978). Calcium-induced release of calcium from the sarcoplasmic reticulum of skinned cells from adult human, dog, cat, rabbit, rat, and frog hearts and from fetal and new-born rat ventricles. *Ann. N.Y. Acad. Sci.*, 307, 491-522.

Fozzard, H. A. (1977). Heart: Excitation-contraction coupling. *Ann. Rev. Physiol.*, 39, 201-220.

Hungerford, R. T. and Lindenmayer, G. E. (1981). Two components of Na-dependent Ca uptake by cardiac sarcolemma. *Fed. Proc.*, 40, 578.

Jones, L. R., Besch, H. R., Jr., Fleming, J. W., McCannaughey, M. M., and Watanabe, A. M. (1979). Separation of vesicles of cardiac sarcolemma from vesicles of cardiac sarcoplasmic reticulum. *J. Biol. Chem.*, 254, 530-539.

Katz, A. M. (1977). "Physiology of the Heart." Raven Press, N.Y.

Miyamoto, H. and Racker, E. (1980). Solubilization and partial purification of the Ca^{2+}/Na^{+} antiporter from the plasma membrane of bovine heart. *J. Biol. Chem.*, 255, 2656-2658.

Mullins, L. J. (1977). A mechanism for Na/Ca transport. *J. Gen. Physiol.*, 70, 681-695.

Philipson, K. D. and Nishimoto, A. Y. (1980). Na^{+}-Ca^{2+} exchange is affected by membrane potential in cardiac sarcolemma vesicles. *J. Biol. Chem.*, 255, 6880-6882.

Pitts, B. J. R. (1979). Stoichiometry of sodium-calcium exchange in cardiac sarcolemmal vesicles. *J. Biol. Chem.*, 254, 6232-6235.

Reeves, J. P. and Sutko, J. L. (1979). Sodium-calcium ion exchange in cardiac membrane vesicles. *Proc. Natl. Acad. Sci. USA*, 76, 590-594.

Reeves, J. P. and Sutko, J. L. (1980). Sodium-calcium exchange activity generates a current in cardiac membrane vesicles. *Science*, 208, 1461-1464.

Reuter, H. (1979). Properties of two inward membrane currents in the heart. *Ann. Rev. Physiol.*, 41, 413-424.

Van Alstyne, E., Burch, R. M., Knickelbein, R. G., Webb, J. G., Hungerford, R. T., Gower, E. J., Poe, S. L., and Lindenmayer, G. E. (1980). Isolation of sealed vesicles highly enriched with sarcolemma markers from canine left ventricle. *Biochim. Biophys. Acta*, 602, 131-143.

Van Winkle, W. B. and Schwartz, A. (1976). Ions and inotropy. *Ann. Rev. Physiol.*, 38, 247-272.

KINETICS OF SERUM PROTEIN SECRETION BY CULTURED HEPATOMA CELLS: EVIDENCE FOR MULTIPLE SECRETORY PATHWAYS[1]

Barry E. Ledford and Donna F. Davis

Department of Biochemistry
Medical University of South Carolina
171 Ashley Avenue
Charleston, South Carolina 29425

[1]This investigation was supported in part by Grant CA 17037 from the National Cancer Institute and in part by the South Carolina Biomedical Research Appropriation for 1979-80.

I. INTRODUCTION

The intracellular processing of integral membrane proteins, secretory proteins, and certain organeller proteins requires signals which segregate these proteins to their respective destinations (Blobel, 1980). The most obvious of these is the presence of a N-terminal sequence of amino acids which is responsible for the interaction of the nascent polypeptide with the membrane of the RER (Blobel and Sabatini, 1971; Milstein et al., 1972). The amino terminus, by virtue of its hydrophobic character, might interact directly with components of the membrane (Blobel and Dobberstein, 1975a) or might alter the folding of the nascent peptide such that membrane interactive domains are exposed (Wickner, 1979). Once associated with the membrane, the growing polypeptide is directed into or through the membrane. Cotranslational modifications generally include the proteolytic cleavage of the amino terminal extension and the addition of N-linked core oligosaccharides (Campbell and Blobel, 1976; Waechter and Lennarz, 1976). The remarkable aspect of the processing events occurring in the RER is the lack of tissue or even species specificity. Blobel and Dobberstein (1975b) initially demonstrated that dog pancreas membranes would correctly process secretory proteins from a diverse range of species. It would thus appear that in a particular cell all proteins destined for secretion probably utilize the same processing elements in the RER. Is the remainder of the intracellular processing pathway the same for all secretory proteins produced by a cell, particularly a cell secreting several proteins?

The liver is a major secretory organ producing the majority of the plasma proteins (Miller and John, 1970) with the exception of the immunoglobulins (Putnam, 1977) and certain components of the complement system (Müller-Eberhard, 1975). The physical and chemical characteristics of these proteins are diverse, including glycoproteins, albumin (a nonglycosylated protein), single polypeptides, and multimeric proteins. Morgan and Peters (1971) initially observed that albumin was transported out of liver cells faster than transferrin. This has been confirmed by Schreiber et al. (1979). Strous and Lodish (1980) have further shown in VSV-infected hepatoma cells that the processing of the integral membrane glycoprotein G and transferrin differ in the rate of oligosaccharide maturation. These differences in rates of processing and secretion within a single cell indicate that there is more than one secretory pathway in liver cells.

The cell line Hepa, derived from the mouse hepatoma BW7756 (Bernhard et al., 1973), continues to secrete the serum proteins albumin, transferrin, and α-fetoprotein. We have studied the rates of intracellular processing of these proteins using either pulse-chase or variable-pulse labeling techniques. The results indicate that the kinetics of albumin secretion differ from those of transferrin and α-fetoprotein. The implications of these results are that the proteins are segregated to a degree during intracellular processing. Potential sorting signals and sites of segregation are discussed, and parallels are drawn between secretory processing in the liver and in the parathyroid gland.

II. MATERIALS AND METHODS

Growth of Cells

Hepa cells were grown in Dulbecco's Modified Eagle's medium (DMEM) supplemented with 2% fetal calf serum, penicillin (100 units/ml), and streptomycin (100 μg/ml). All experiments were carried out in 35 mm Petri dishes in media made from stock solution of amino acids, vitamins, and salts. Experimental media were similar to DMEM except that leucine and methionine concentrations were reduced in labeling studies. The medium for leucine labeling contained 35 mg leucine/liter (Ledford et al., 1977), whereas, the methionine labeling medium contained 3 mg methionine/liter. All cultures were maintained in a humidified atmosphere of 5% CO_2-95% air. Cell numbers were determined by releasing the cells in trypsin: EDTA and counting in a hemacytometer.

Cell-Free Translation

Preparation of polyribosomes for either RNA extraction or polyribosome readout was done essentially as described by Fioretti et al. (1979) except that heparin and cycloheximide were eliminated from the homogenization and gradient buffers. Total Hepa RNA was prepared from polyribosomes according to the method described by Deeley et al. (1977). The final dried RNA pellet was dissolved in diethylpyrocarbonate treated distilled water.

Cell-free translation was carried out in a rabbit reticulocyte derived system described by Pelham and Jackson (1976). Aurin tricarboxylic acid (10^{-4}M) was included to prevent initiation when polyribosomes were readout (Lodish et al., 1971). Translation was allowed to procede for 1 hr at 30°C. The mixture was then centrifuged at 50,000 rpm in a Beckman type 65 rotor. The supernatant was made 0.02% in sodium azide before immunoprecipitation of albumin precursors.

Immunoprecipitation

The preparation of antisera directed against the mouse serum protein albumin, α-fetoprotein, and transferrin has been described elsewhere (Allen and Ledford, 1977). The titer and specificity of all antisera were routinely monitered. All immunoprecipitations were carried out under conditions of antibody excess. Immunoprecipitates were washed five times with 2 ml of 0.9% NaCl. The final precipitate was dissolved in 1 ml of 0.1 N NaOH and mixed with liquid scintillation cocktail.

Radioimmunoassay

Antigens to be used in radioimmunoassays were dialyzed against 0.2 M phosphate buffer, pH 7.2. Enzymobeads (Biorad) were used in the iodination reaction according to the instructions provided by the supplier. 100 μg of protein was iodinated with 2 mCi of Na ^{125}I. The protein was separated from the reaction mixture by chromatography on G-25 Sephadex (Pharmacia). The resulting specific activities were approximately 5 x 10^6 cpm/μg.

The rabbit antisera directed against the mouse serum proteins were diluted such that less than 30% of the iodinated antigen bound in the absence of competing antigen. Incubation of diluted rabbit antisera with standards or unknowns was at 4°C for 24 hr. Carrier rabbit serum and goat anti-rabbit IgG was then added and incubated at 4°C for 24 hr.

The immunoprecipitates were washed five times with 0.9% NaCl, dissolved in 1 ml of 0.1 N NaOH, and counted by liquid scintillation. Assay conditions were adjusted such that a 50% reduction in labeled antigen occurred at 20, 10, and 5 ng of competing albumin, α-fetoprotein, or transferrin, respectively. Standard curves were transformed to linear presentations, fitted by linear regression methods, and used to calculate amount of antigen in samples.

Gel Electrophoresis

The amount of carrier antigen was reduced to 5 μg for immunoprecipitates to be analyzed by polyacrylamide gel electrophoresis. Samples were electrophoresed in 7.5% SDS polyacrylamide slab gels made according to Laemmli (1970). Immunoprecipitates were dissolved in 250 μl of sample buffer (O'Farrell, 1975) and heated to 100°C for 10 min. After electrophoresis, gels were impregnated with PPO (Bonner and Laskey, 1974), dried, and fluorographed on preflashed Kodak XRA-5 film (Laskey and Mills, 1975). Fluorographs were scanned using a Joyce-Loebl densitometer.

Labeling of Hepa Proteins

The rate of appearance of ^{3}H-leucine-labeled albumin in the culture medium was measured as described previously (Ledford et al., 1977). The specific activity of leucine in the medium, the cell number, and the leucine content of mouse albumin were then used to calculate a cellular rate of secretion expressed in terms of molecules of albumin/cell/min.

The fates of intracellular precursors to Hepa secretory proteins were followed as a function of time in either pulse-chase or variable pulse labeling experiments. The cells were harvested in 1 ml of polysome buffer (250 mM ribonuclease-free sucrose, 5 mM $MgCl_2$, 25 mM NaCl, 40 mg/ml heparin, 1% Triton X-100, 1% sodium deoxycholate, 20 mM Tris, pH 7.4) and homogenized in a glass-glass homogenizer. The homogenate was centrifuged at 10,000 g for 10 min. The supernatant was then centrifuged at 50,000 rpm in a type 65 rotor (Beckman) for 1 hr. Proteins were immunoprecipitated from the resulting supernatant as described above.

III. RESULTS

The rates of serum protein secretion by Hepa cells can be quantitated either directly by radioimmunoassay or by measurement of labeled amino acid incorporation. The use of labeled amino acids in such studies is generally preferred since estimates of cellular lifetime can be made and since it permits the identification of secretory intermediates using electrophoretic techniques. The quantitation of secretion rates from isotope incorporation data requires that one know the intracellular specific activity of the precursor. Linear incorporation rates are indicative of steady-state precursor pool specific activity. In the case of Hepa cells, the medium volume is approximately 100-fold greater than the cellular volume. Thus, at equilibrium the specific activity in the cells should closely approximate that of the medium. Rather than attempt to measure the intracellular specific activity directly and still be uncertain about potential variations among microenvironments, we compared radioimmunoassay measurements with incorporation-derived measurements of secretion rates. Figure 1 shows the time-dependent appearance of mouse albumin in the culture medium. The amoung of ^3H-leucine incorporated into albumin was converted to µg of albumin from the specific activity of leucine in the medium and the leucine content of mouse albumin (Popp et al., 1966). The values obtained by radioimmunoassay and leucine incorporation were corrected for the fraction of the total volume sampled and for albumin removed in previous samples. ^3H-labeled albumin begins appearing in the medium approximately one hour after the addition of ^3H-leucine to the medium. Albumin measured by radioimmunoassay increases linearly from the beginning of the experiment. The initial value represents albumin which was secreted during the 1 hr acclimation period following the medium change. The two accumulation curves are nearly parallel indicating that incorporation-derived measurements are consistent with the assumption that the specific activity of leucine in the medium

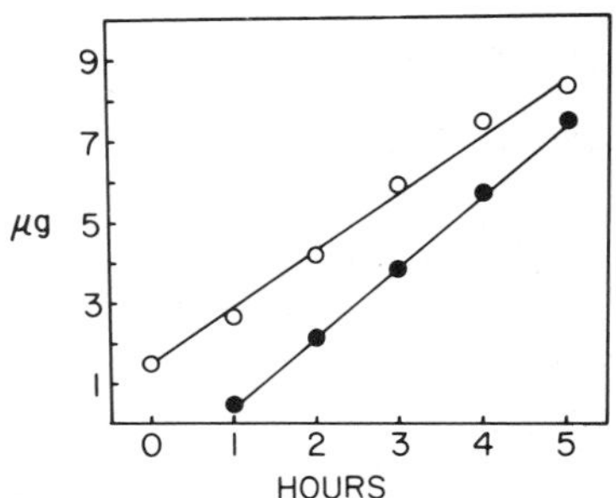

Fig. 1. Albumin secretion rates measured by radioimmunoassay and by ^{3}H-leucine incorporation. The media from companion 35 mm cultures of Hepa cells were removed and replaced with 1 ml of labeling medium containing 35 mg leucine/liter. Cultures were allowed to acclimate for 1 hr. At time 0, 50 μCi of ^{3}H-leucine was added to one culture. 25 μl aliquots of the culture media were taken at hourly intervals and assayed for mouse albumin either by radioimmunoassay or by direct immunoprecipitation of labeled albumin. Values were then corrected to represent the total amount of albumin in the media. Radioimmunoassay, O--O; ^{3}H-leucine incorporation ●--●.

closely approximates that in the cell. The time required for the appearance of labeled albumin in the medium is an estimate of the intracellular residence time since actual synthesis only requires 3 min (Hamman et al., 1980).

The dynamics of the intracellular processing of serum proteins by Hepa was investigated using a pulse-chase labeling approach. The cells were pulse-labeled for 10 min with ^{35}S-methionine. The medium was then replaced with medium containing a 100-fold higher concentration of methionine. At various times afterward, individual proteins were immunoprecipitated from either detergent-solubilized cells or from culture media. Figure 2 shows the disappearance of labeled albumin from the cells and its appearance in the culture medium. The data show that 80% of newly synthesized albumin exits the cells within 60 min. The remainder appears to persist in the cells for an extended period of time. In other experiments (data not shown) a 16 hr chase did not further reduce the residual labeled albumin within the cells. Two potential explanations for this apparent retention of albumin were tested. First, if

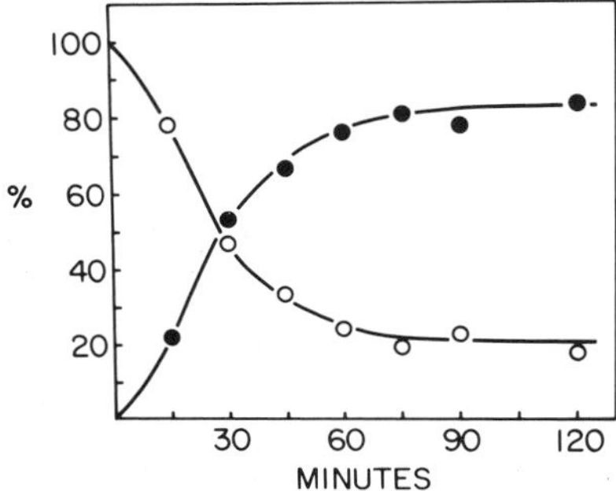

Fig. 2. Pulse-chase labeling of intracellular and extracellular albumin. Growth medium was removed from 35 mm cultures of Hepa cells and replaced with 1.0 ml of methionine deficient medium. After a 1 hr equilibration period, 10 μCi of ^{35}S-methionine was added to each culture. Incorporation was allowed to proceed for 10 min. The medium was then removed, cultures were washed with 1 ml growth medium, and 1 ml of growth medium containing 300 mg methionine/liter.

At the end of the chase period, the medium was removed and the cells were harvested in polysome buffer and homogenized. Albumin was immunoprecipitated from the postribosomal supernatant and the medium as described under Materials and Methods. The values shown are expressed as the percentage of total (extracellular + intracellular) albumin. Intracellular albumin, 0--0; Extracellular albumin, ●--● .

there was a lack of immunospecificity, the residual label might be contaminating protein rather than albumin. This was examined by displaying the immunoprecipitates on SDS-polyacrylamide gels. As shown in Figure 3, the immunoprecipitated label is associated with a single species having an apparent molecular weight approximately that of albumin. The persistent label is only associated with albumin eliminating a lack of immunospecificity as an explanation. The second possibility considered was that the specific activity of the amino acid pool had not been sufficiently reduced to effectively terminate the incorporation of label into protein. Thus, the persistence of labeled albumin in the cells might be due to continued incorporation of label. The effectiveness of the chase was investigated by measuring the amount of label

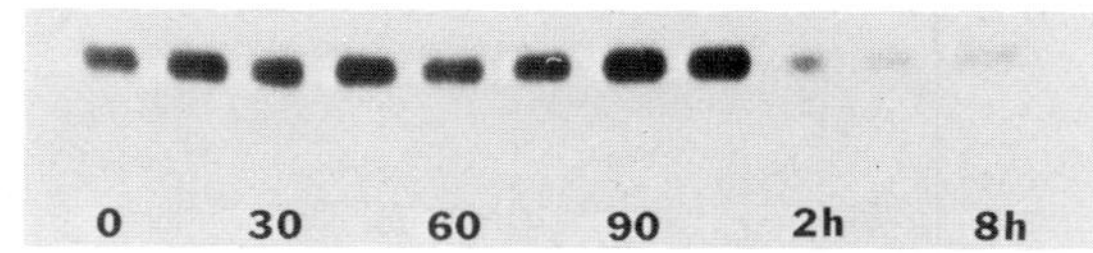

Fig. 3. Electrophoretic analysis of albumin immunoprecipitated after pulse-chase labeling of Hepa cells. Cultures were labeled as described for Figure 2, except 200 μCi of ^{35}S-methionine was used. Immunoprecipitates were dissolved in SDS sample buffer, electrophoresed and fluorographed as described under Materials and Methods. The chase times are shown.

incorporated into total protein during the experiment. Table 1 shows the amount of TCA-precipitable label at various chase times. These data show that the chase very effectively blocked any further incorporation of label. Thus, the labeled albumin remaining in the cells must have been synthesized during the pulse period.

The secretion dynamics of α-fetoprotein and transferrin were measured simultaneously with those shown for albumin in Figure 2. Figure 4 shows the time dependent appearances of these proteins in the medium. Transferrin and α-fetoprotein exit the cells more slowly than albumin and are retained to a greater degree than is albumin. The specificities of the immunoprecipitations of transferrin and α-fetoprotein were examined and found to be equivalent to that shown for albumin (data not shown). Thus, there is a selective retention of some secretory proteins within Hepa cells.

The dynamics of protein secretion were further examined using a variable pulse-labeling approach. Cells were pulse-labeled for various times with ^{3}H-leucine. Proteins were then immunoprecipitated from either detergent-solubilized cells or from the culture media. The time dependent labeling of intracellular and extracellular albumin is shown in Figure 5. A steady-state level of labeled albumin in the cells is achieved by approximately 1-2 hr at which time the rate of labeled albumin accumulation in the medium becomes linear. Since albumin, α-fetoprotein, and transferrin are not secreted at equivalent rates by Hepa cells, it is not possible to compare such labeling experiments directly. However, if the data are expressed as the ratio of the extracellular label to the intracellular label, the time dependence of these ratios is directly comparable. Figure 6 shows the time dependent distribution of albumin, α-fetoprotein, and transferrin. The rate at which the

Table 1. Hepa cell cultures were pulse-chase labeled with ^{35}S-methionine as described in Figure 2. Total protein was precipitated from 100 μl aliquots of the post-ribosomal supernatants by addition of 1 ml of 10% trichloroacetic acid (TCA). Precipitation mixtures were heated to 100°C for 10 min. The precipitated proteins were collected by centrifugation and washed 5 times with 1 ml of 5% TCA. Final precipitates were dissolved in 1 ml of 0.1 N NaOH and counted by liquid scintillation.

Pulse-Chase Labeling of Total Hepa Protein with [^{35}S]-Methionine

Chase Time (minutes)	TCA-Precipitable Radioactivity (cpm x 10^{-5})
0	2.8
15	2.9
30	2.9
45	2.9
60	2.5
75	3.1
90	2.8

distributions of α-fetoprotein and transferrin change is slower than that of albumin by approximately a factor of three. This is a result of a higher degree of intracellular retention of α-fetoprotein and transferrin.

Rates of secretion of albumin, α-fetoprotein, and transferrin were measured directly by radioimmunoassay. Steady-state levels of these proteins within the cells were also measured by radioimmunoassay. Figure 7 shows the time dependent accumulation of the proteins in the culture medium. Secretion rates were calculated from the slopes of the resultant lines. The intracellular content of each protein can then be calculated in terms of synthesis time. These results are shown in Table 2. The intracellular contents of α-fetoprotein and transferrin are higher

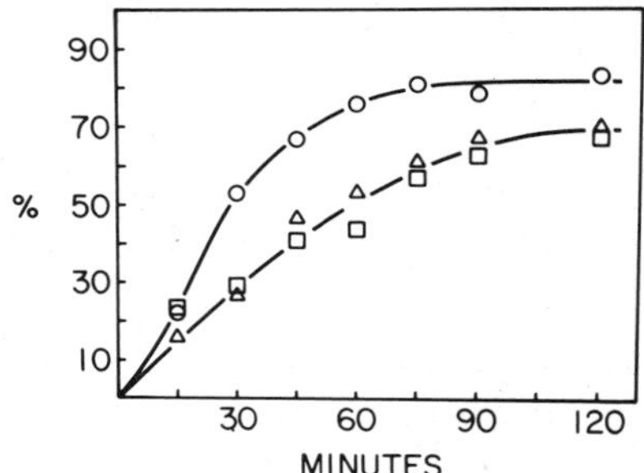

Fig. 4. Chase of pulse-labeled proteins into the culture medium. The labeling and immunoprecipitation are described for Figure 2. The values shown are the percentage of total (cellular + medium) immunoprecipitable label in the medium. Albumin, O--O; α-fetoprotein, □--□ ; transferrin, △--△ .

than that of albumin. Thus, labeling approaches and direct radioimmunoassay of secretory proteins leads to the same conclusions that α-fetoprotein and transferrin are retained within Hepa cells to a greater extent than is albumin.

One potential reason for the retention of hepatic secretory proteins is nonmembrane associated synthesis. Thus, if a portion of total synthesis occurred in the cytoplasm rather than on the RER, the completed protein would remain free in the cytoplasm. Yap et al. (1977) have now shown that under normal conditions that this does not occur in liver. However, in the livers of starved rats, a portion of the albumin mRNA is free in the cytoplasm (Yap et al., 1978), albeit nonfunctional. If cytoplasmic synthesis of secretory proteins was occurring then the cytoplasm would be expected to be identical to the protein synthesized in a mRNA primed cell-free protein synthesis system. That is, it should possess an amino terminal extension and not be glycosylated. In the case of the glycoproteins, glycosylation, and proteolytic cleavage of the N-terminal extension have offsetting effects on mass. Albumin, however, is not a glycoprotein. Thus a cytoplasmically-synthesized albumin should be approximately 2000 daltons larger than the membrane associated product (Strauss et al., 1978). This was examined by comparing the cell-free translation product with intracellular albumin. Figure 8 shows that on SDS-

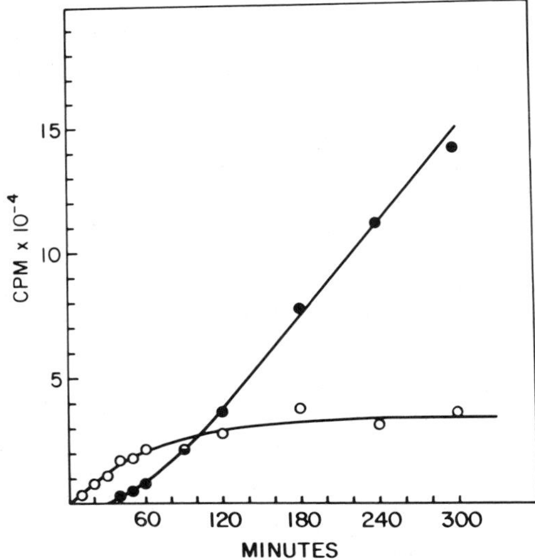

Fig. 5. Time dependent labeling of intracellular and extracellular albumin. Growth medium was removed from 35 mm cultures of Hepa cells and replaced with 1 ml of medium containing 35 mg leucine/liter. After acclimating for 1 hr, 25 μCi of ^{3}H-leucine (Amersham, 76 Ci/m mole) was added. At the times shown, the medium was removed, the cells were harvested in 1 ml polysome buffer, and homogenized. Albumin was immunoprecipitated from the postmitochondrial supernatant and from the medium as described in Material and Methods. Intracellular albumin, O--O; Extracellular albumin, ●--● .

polyacrylamide gel electrophoresis the cell-free product is larger than intracellular albumin. Since multiple forms of albumin were not observed in Hepa it must be concluded that cytoplasmic synthesis of albumin is not responsible for intracellular retention of the protein.

IV. DISCUSSION

The postranslational modification and intracellular movement of secretory proteins occurs in vesicles (Palade, 1975). Thus, any consideration of multiple secretory pathways within cells must include some degree of protein segregation among vesicles. The simplest case would involve the segregation of proteins for their

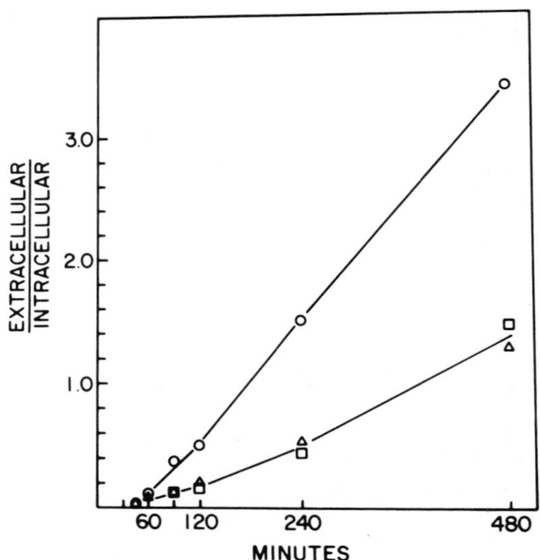

Fig. 6. Time dependent distribution of Hepa secretory proteins. Cultures of Hepa cells were labeled, and secretory proteins immunoprecipitated as described in Figure 5. The values shown are the ratios of extracellular (medium) to intracellular radioactivity. Albumin, O--O; α-fetoprotein, □--□ ; transferrin, △--△ .

entire intracellular residence time. Such segregation would require sorting of the proteins at the RER. Once sorted into transition elements the protein's intracellular movement would be completely governed by the movement of the vesicle. Convergent pathways would represent a variation on this scheme. Sorting would still occur at the RER; however, somewhere during intracellular transit vesicles would fuse, perhaps at the Golgi. Alternatively, divergent pathways would require the sorting of proteins somewhere other than the RER, the Golgi being a likely site.

The previous evidence supporting the existence of multiple secretory pathways in liver (Morgan and Peters, 1971; Schreiber et al., 1979) has been based on intracellular residence time differences between albumin and transferrin. This is consistent with the segregation of albumin and transferrin into different

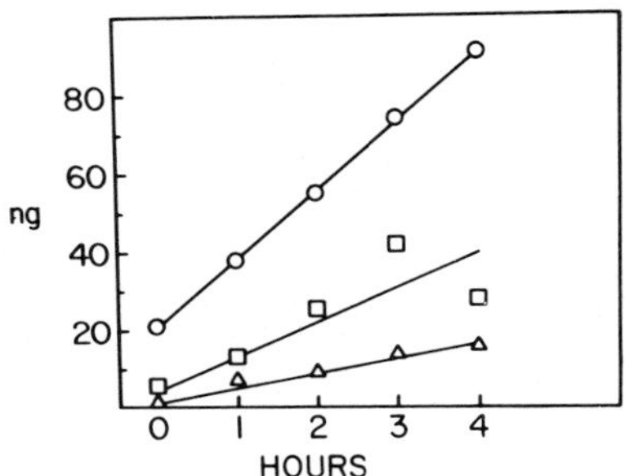

Fig. 7. Rates of serum protein secretion measured by radioimmunoassay. The culture medium was removed from three companion 35 mm cultures of Hepa and replaced with 1 ml of fresh medium/culture. Cultures were allowed to accumulate for 1 hr. 25 µl aliquots of the media were taken at the indicated times. The amounts of albumin, α-fetoprotein, or transferrin were analyzed by radioimmunoassay. Values were corrected for the fraction of the total medium sampled and for protein removed in previous samples, to represent the total amount of each protein in the medium. Albumin, 0--0; α-fetoprotein, □ -- □ ; transferrin Δ -- Δ .

vesicles. The data of Strous and Lodish (1980) on the rates of oligosaccharide chain maturation and intracellular residence times indicate that in VSV-infected hepatoma cells segregation occurs before the Golgi.

The data presented here indicate that multiple secretory pathways exist in Hepa cells. This conclusion is based on two independent observations. First, the kinetics of albumin secretion (Fig. 4) differ from those of α-fetoprotein and transferrin. Secondly, α-fetoprotein and transferrin are retained with Hepa cells to a greater degree than is albumin. Since the retained albumin is not preproalbumin, it must have been synthesized in association with membranes and thus be in vesicles. Varying degrees of retention then indicate that some segregation of proteins among vesicles occurs in Hepa cells. The different kinetics of secretion are also consistent with vesicular segregation. That is, if secretory proteins have different intracellular residence times they cannot be traveling in the same vesicle. The data presented here can be interpreted in

Table 2. Radioimmunoassay of Hepa secretory proteins. The rates of accumulation of Hepa secretory proteins in the culture medium was measured by radioimmunoassay as described for Figure 1. After the samples for determination of the secretion rates (A) were taken, the cells were harvested in polysome buffer and homogenized. The intracellular content (B) of secretory proteins was determined by radioimmunoassay of the postribosomal supernatant. The intracellular contents expressed in terms of hours of synthesis was calculated as the ratio of the secretion rate and the intracellular content in ng.

	ALB	AFP	Tf
A. SECRETION RATE (ng/hr/dish)	17.6	3.88	7.31
B. INTRACELLULAR	50.2 ± 2.7	14.1 ± 1.2	24.9 ± 4.0
C. INTRACELLULAR CONTENTS, A/B (hrs. of Synthesis)	2.85	3.63	3.40

terms of different vesicles. They cannot, however, be interpreted to indicate that proteins are within the same vesicles. Thus, α-fetoprotein and transferrin while exhibiting very similar secretion kinetics and levels of intracellular retention are not necessarily within the same vesicle.

The segregation of proteins among vesicles requires signals which are intrinsic to the protein. Although albumin and α-fetoprotein possess a high degree of sequence homology (Liao et al., 1980), they are sorted differently in Hepa cells. The common structural feature of α-fetoprotein and transferrin is the presence of N-linked oligosaccharides. These are probably not sorting signals since Strous and Lodish (1980) have demonstrated that the glycosylated G protein of VSV is processed differently than transferrin in rat hepatoma cells. A potential sorting signal is the propiece of albumin. This basic hexapeptide is cleaved postranslationally by a cathepsin D-like enzyme prior to secretion (Judah and Quinn, 1978).

Multiple secretory pathways have also been proposed for proteins secreted by the parathyroid gland (Morrissey and Cohn, 1978). Parthyroid hormone, like albumin, is not glycosylated

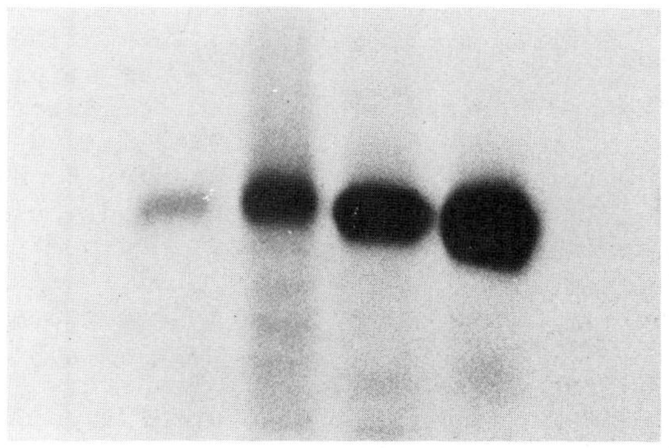

Fig. 8. Comparison of albumin synthesized in Hepa cells with albumin synthesized in Hepa cells with albumin synthesized in a reticulocyte lysate. Polyribosomes from Hepa cells were purified by centrifugation in sucrose density gradients and used for RNA purification or cell-free readout as described in Materials and Methods. Aurin tricarboxylic acid (1×10^{-4} M) was included in the reticulocyte lysate to inhibit initiation when ribosomes were readout. After translation, lysates were centrifuged at 50,000 rpm for 1 hr in a type 65 rotor (Beckman) to remove nascent proteins. Immunoprecipitation of albumin from the supernatant was accomplished by the same procedure used for postribosomal supernatants from Hepa cells. Immunoprecipitates were displayed on SDS polyacrylamide slab gels and fluorographed as described before. Lane 1, Hepa RNA; Lane 2, Hepa polyribosomes; Lane 3, intracellular albumin; Lane 4, extracellular albumin.

and exists in the cell with a basic N-terminal hexapeptide extension. The parathyroid gland also secretes glycoproteins. Glycoprotein secretion and parathyroid secretion do not respond identically to changes in external calcium concentrations. Such a differential response can most readily be accounted for by segregating the proteins into different secretory vesicles (Morrissey and Cohn, 1978). The apparent similarities between the intracellular processing of hepatic secretory proteins and parathyroid secretory proteins suggests that the propiece of albumin and parathyroid hormone may function in the intracellular sorting of these proteins.

V. REFERENCES

Allen, R. P. and Ledford, B. E. (1977). The influence of antisera specific for α-fetoprotein and mouse serum albumin on the viability and protein synthesis of cultured mouse hepatoma cells. *Cancer Res.*, 37, 696-701.

Bernhard, H. P., Darlington, G. J., and Ruddle, F. H. (1973). Expression of liver phenotypes in cultured mouse hepatoma cells: synthesis and secretion of serum albumin. *Dev. Biol.*, 35, 83-96.

Blobel, G. (1980). Intracellular protein topogenesis. *Proc. Natl. Acad. Sci.*, 77, 1496-1500.

Blobel, G and Dobberstein, B. (1975a). Transfer of proteins across membranes I. Presence of proteolytically processed and unprocessed nascent immunoglobulin light chains on membrane-bound ribosomes of murine myeloma. *J. Cell Biol.*, 67, 835-851.

Blobel, G and Dobberstein, B. (1975b). Transfer of proteins across membranes II. Reconstitution of functional rough microsomes from heterologous components. *J. Cell Biol.*, 67, 852-862.

Blobel, G. and Sabatini, D. D. (1971). Ribosome-membrane interaction in eukaryotic cells. In "Biomembranes". vol. 2 (L. A. Manson, ed.) pp. 193-195. Plenum Press, N.Y.

Bonner, W. M. and Laskey, R. A. (1974). A film detection method for tritium-labelled proteins and nucleic acids in polyacrylamide gels. *Eur. J. Biochem.*, 46, 83-88.

Campbell, P. N. and Blobel, G. (1976). The role of organelles in the chemical modification of the primary translation products of secretory proteins. *FEBS Letters*, 72, 215-226.

Deeley, R. G., Gordon, J. I., Burns, A. T. H., Mullinix, K. P., Binastein, M., and Godberger, R. F. (1977). Primary activation of the vitellogenin gene in the rooster. *J. Biol. Chem.*, 252, 8310-8319.

Fioretti, W. C., Davis, D. F., and Ledford, B. E. (1979). Polyribosome size analysis: measurement of number-average polyribosome sizes. *Biochim. Biophys. Acta*, 564, 79-89.

Hamman, H. C., Simpson, J. A., and Ledford, B. E. (1980). Effects of cyclic AMP on the kinetics of serum protein synthesis in cultured mouse hepatoma cells. *Arch. Biochem. Biophys.*, 204, 277-287.

Judah, J. D. and Quinn, P. S. (1978). Calcium ion-dependent vesicle fusion in the conversion of proalbumin to albumin. *Nature*, 271, 384-385.

Laemmli, U. K. (1970). Cleavage of structural proteins during the assembly of the head of bacteriophage T4. *Nature*, 227, 680-685.

Laskey, R. A. and Mills, A. D. (1975). Quantitative film detection of ^{3}H and ^{14}C in polyacrylamide gels by fluorography. *Eur. J. Biochem.*, 56, 335-341.

Ledford, B. E., Warner, R. W., and Cochran, R. A. (1977). Albumin synthesis in cultured hepatoma cells: regulation by essential amino acids. *Biochim. Biophys. Acta*, 475, 90-95.

Liao, W. S. L., Hamilton, R. W., and Taylor, J. M. (1980). Amino acid sequence homology between rat α-fetoprotein and albumin at the COOH-terminal regions. *J. Biol. Chem.*, 255, 8046-8049.

Lodish, H. F., Housman, D., and Jacobsen, M. (1971). Initiation of hemoglobin synthesis. Specific inhibition by antibiotics and bacteriophage ribonucleic acid. *Biochemistry*, 10, 2348-2356.

Miller, L. L. and John, D. W. (1970). Nutritional, hormonal, and temporal factors regulating net plasma protein biosynthesis in the isolated perfused rat liver. Effects of feeding or fasting liver donors and of supplementation with amino acids, insulin, cortisol, and growth hormone. In "Plasma Protein Metabolism". (M. A. Rothschild and Waldman, T., eds.) pp. 207-222. Academic Press, N.Y.

Milstein, G., Brownlee, G. G., Harrison, K. M., and Mathews, M. B. (1972). A possible precursor of immunoglobulin light chains. *Nature New Biol.*, 239, 117-120.

Morgan, E. H. and Peters, T. (1971). Intracellular aspects of transferrin synthesis and secretion in the rat. *J. Biol. Chem.*, 246, 3508-3511.

Morrissey, J. J. and Cohn, D. V. (1978). The effects of calcium and magnesium on the secretion of parathormone and parathyroid secretory protein by isolated porcine parathyroid cells. *Endo.*, 103, 2081-2090.

Müller-Eberhard, H. J. (1975). Complement. *Ann. Rev. Biochem.*, 44, 697-724.

O'Farrell, P. H. (1975). High resolution two-dimensional electrophoresis of proteins. *J. Biol. Chem.*, 250, 4007-4021.

Palade, G. (1975). Intracellular aspects of the process of protein synthesis. *Science*, 189, 347-358.

Pelham, H. R. B. and Jackson, R. J. (1976). An efficient mRNA-dependent translation system from reticulocyte lysates. *Eur. J. Biochem.*, 67, 247-256.

Popp, R. A., Heddle, J. G., Canning, R. E., and Allen, R. C. (1966). Some physical and chemical properties of albumin esterase and albumin from mouse serum. *Biochim. Biophys. Acta*, 115, 113-120.

Putnam, F. W. (1977). Immunoglobulins 1. Structure. In "The Plasma Proteins" 2nd edition. (F. W. Putnam, ed.) Vol. III, pp 1-153. Acadmemic Press, N.Y.

Schreiber, G., Dryburgh, H., Millership, A., Matsuda, Y., Inglis, A., Phillips, J., Edwards, K., and Maggs, K. (1979). The synthesis and secretion of rat transferrin. *J. Biol. Chem.*, 254, 12013-12019.

Strauss, A. W., Bennett, C. D., Donohue, A. M. Rodkey, J. A., and Alberts, A. W. (1977). Rat liver pre-proalbumin: complete amino acid sequence of the pre-piece. Analysis of the direct translocation product of albumin messenger RNA. *J. Biol. Chem.*, 252, 6846-6855.

Strous, G. J. A. M. and Lodish, H. F. (1980). Intracellular transport of secretory and membrane proteins in hepatoma cells infected by VSV. *Cell*, 22, 709-717.

Waechter, C. J. and Lennarz, W. J. (1976). The role of polyprenol-linked sugars in glycoprotein synthesis. *Ann. Rev. Biochem.*, 45, 95-112.

Wickner, W. (1979). The assembly of proteins into biological membranes: the membrane trigger hypothesis. *Ann. Rev. Biochem.*, 48, 23-45.

Yap, S. H., Strair, R. K., and Shafritz, D. A. (1977). Distribution of rat liver albumin mRNA membrane-bound and free polyribosomes as determined by molecular hybridization. *Proc. Natl. Acad. Sci.*, 74, 5397-5401.

Yap, S. H., Strair, R. K., and Shafritz, D. A. (1978). Effect of a short term fast on the distribution of cytoplasmic albumin messenger ribonucleic acid in rat liver. Evidence for formation of free albumin messenger ribonucleoprotein particles. *J. Biol. Chem.*, 253, 4944-4950

MULTIPLE ANTIGEN BINDING CELLS AND B-LYMPHOCYTE MATURATION

Dominick Deluca

Department of Biochemistry
Medical University of South Carolina
171 Ashley Avenue
Charleston, South Carolina 29425

I. INTRODUCTION

Since the advent of the clonal selection theory (Burnet, 1959) it has been generally accepted that the phenotypic expression of receptor specificities in the immune system is restricted at the level of the lymphocyte, i.e. one receptor specificity per cell, no matter what the differentiation state of the cell. Adoptive transfer experiments (Playfair et al., 1965; Kennedy et al., 1966; Shearer et al., 1969) tended to confirm this notion in that only a small proportion of cells (~ 1 in 10^5) has the capacity to respond to a given antigen. These systems depend on the expression of an immune response by B cells, a process that is much more complicated than the simple combination of B cell receptors with antigen. Early studies of antigen binding (Naor and Sulitzeanu, 1967; Byrt and Ada, 1969) done with low concentrations of antigen that did not give a plateau frequency of antigen binding cells (ABCs), showed a frequency of ABCs similar to that found for responsive B cells in adoptive transfer systems. Other ABC studies (Urbain-Vansanten et al., 1974; Rutishauser et al., 1972; Lawrence et al., 1973; Julius and Herzenberg, 1974; Donald et al., 1974; Cooper et al., 1972; Julius et al., 1976; Dwyer and MacKay, 1972; Ada, 1970) in which saturating amounts of antigen were used yielded many more ABCs (1 in 10^2 - 10^3 in the unimmunized spleen). These latter data were extended to include the binding of two non-cross-reacting antigens to show that the high frequencies of ABCs were not due to binding of both antigens to the same "sticky" population of cells (Dwyer and MacKay, 1972). However, these high frequencies of ABCs (1 in 10^2 - 10^3) suggest that some cells capable of binding more than one antigen should exist. Some workers using fluorochromated antigens (Urbain-Vansanten et al., 1974; DeLuca, unpubl.) have shown directly that while most ABCs bind only one of a given pair of antigens, a small number bind both antigens.

II. CLONAL NATURE OF ANTIGEN BINDING

Earlier studies utilizing fluorescent protein antigens had established that a small proportion of cells with lymphoid morphology in unimmunized mouse spleen and bone marrow were capable of binding more than one antigen concurrently by non-crossreactive Ig receptors (DeLuca et al., 1975). The frequency of ABCs for any given antigen could be predicted by determining the product of the frequencies of ABC for each antigen alone among the total Ig-bearing cell population (nearly all ABCs also bear surface Ig)

(DeLuca, unpubl.). Bone marrow lymphocytes (Table 1) have a higher proportion of ABCs for several protein antigens among Ig-bearing lymphocytes, and, therefore, more double ABCs than spleen (Table 2); even though the latter organ has a much higher proportion of ABCs and Ig-bearing cells than the bone marrow population.

The lymphoid nature of multiple ABCs has been established through the use of pure populations derived from B lymphocyte colony-forming cells grown in agar (Table 3). Colony-forming lymphocytes from spleen and bone marrow give the same proportions of single and double ABCs found among the Ig-bearing cell population (determined by double immunofluorescence with antigen and anti-Ig) taken from the parent organ, i.e., bone marrow colony-derived B lymphocytes have more double ABCs than spleen colony-derived b lymphocytes. Studies utilizing whole, single colonies of bone marrow-derived B lymphocytes indicate that the frequency of single and double ABCs from the population as a whole is directly reflected in the frequency of single and double colonies, a direct test of the clonal nature of antigen-binding capacity (Table 3). No "mixed" ABC colonies, i.e., colonies with some cells binding one antigen and other cells binding the other antigen, were found (DeLuca, unpubl.).

III. SPECIFICITY OF ANTIGEN BINDING

The specificity of antigen binding to these cells was tested by inhibition studies in which antigen binding was inhibited by pre-incubation with anti-immunoglobulin (Table 4). The ability of anti-Ig to inhibit antigen binding was abrogated by prior addition of a pool of mouse myeloma proteins. Since this absorption was performed in liquid phase, complexes of myeloma proteins plus anti-Ig would be expected to form. In spite of the presence of these complexes, inhibition was still prevented, eliminating the possibility that complexes in the original anti-serum were responsible for the inhibition. Excess unlabelled antigen added to the cells prior to the addition of fluorescent labelled antigen prevented the binding of the labelled material except for that binding one might expect from a hapten, e.g., fluorescein, alone (Table 5). Most importantly, the ability of double ABCs to bind one antigen was not inhibited by an excess of the other antigen. The control for this experiment consisted of incubating the cells with an excess of unlabelled antigen after the addition of the same labelled antigen to show that the conditions used actually inhibited the binding of the antigen, especially to double ABCs. These experiments were also performed using spleen cells purified for binding of the NIP-hapten using the method of Hass and Layton (1975), so as to obtain more statistically reliable numbers of double ABCs. The data from that experiment, presented in Table 6,

Table 1. Frequency of "Ring" Ig-bearing cells and antigen binding cells in adult CBA mouse bone marrow.

No. of Experiments	% Ig+[a]	ABC/$10^3 \pm$ SD				Doubles[b]	Doubles Expected[c]
		HSF	KLH	NIP-POL	LDH		
2	5.4 ± 0.8	16 ± 3	-	-	-	16 ± 3	-
2	5.8 ± 2.1	-	-	5 ± 1	-	5 ± 1	-
5	-	9 ± 2	11 + 2	-	-	3 ± 1	1.7
2	-	-		11 + 3	10 ± 3	3 ± 0.3	2.0
5[d]	100	160	196	-	-	53	31
2[d]	100	-	-	196	178	53	34

[a]Stained with IgG fraction of a FITC-labeled polyvalent rabbit anti-mouse immunoglobulin serum.

[b]Indicates double positive cells for each of the reagents used in the experiment shown.

[c]Indicates the frequency of double ABCs expected by the product of the frequencies of total ABCs for each antigen as a fraction of the total Ig-bearing population.

[d]Same data expressed on a per B cell basis, e.e., number of ABC/10^3 normalized to the total marrow cell population which are Ig-bearing lymphoid cells and which also bind antigen.

[-]Indicates not done.

Table 2. Frequency of "Ring" Ig-bearing cells and antigen binding cells in adult CBA spleen.

No. of Experiments	% Ig^{+a}	$\%\mu^{+b}$	ABC/$10^3 \pm$ SD				Doubles[c]	Doubles Expected[d]
			HSF	KLH	NIP-POL	LDH		
1	48.0	-	24	-	-	-	24	-
1	-	34.8	33	-	-	-	33	-
1	38.4	-	-	14	-	-	14	-
2	35.9 + 3.7	-	-	-	24 ± 13	-	24 + 13	-
9	-	-	31 + 7	14 ± 4	-	-	3 ± 1	1.2
3	-	-	-	-	35 ± 4	12 ± 2	3 ± 2	1.2
1	-	-	-	16	-	23	4	1.1
1	-	-	29	-	-	14	4	1.1
g[e]	100	NA	86	38	-	-	8.3	3.3
3[e]	100	NA	-	-	97	33	8.3	3.2
1[e]	100	NA	-	44	-	64	11.1	2.8
1[e]	100	NA	81	-	-	39	11.1	3.1

[a]Stained with the IgG fraction of a FITC-labeled polyvalent rabbit anti-mouse immunoglobulin serum.

[b]Stained with a FITC-labeled polyvalent rabbit anti-μ chain specific serum.

[c]Indicates double positive cells for each of the reagents used in the experiment shown.

[d]Indicates the frequency of double ABCs expected by the product of the frequencies of total ABCs for each antigen among the Ig-positive population.

[e]These figures are calculated on the basis of 35.9% of the spleen cells which bear Ig and also bind antigens.

-Indicates not done.

NA = not applicable.

Table 3. Frequency of "Ring" antigen binding cells among colony derived B lymphocytes and Ig-bearing B cells isolated from spleen and bone marrow.

Cell Type	No. of Experiments	KLH	$ABC/10^3 \pm SD$		
			HSF	Doubles[a]	Doubles Expected[b]
Bone marrow B-lymphocyte colonies	1[c]	137	145	34	20
Pure bone marrow B-lymphocytes	2	172 ± 11	171 ± 56	58 ± 2	29
Whole bone marrow Ig-bearing cells	5	196	160 ± 36	53 ± 18	31
Whole spleen Ig-bearing cells	9	38 ± 11	86 ± 20	8 ± 3	3
Pure spleen B-lymphocytes	2	13 ± 2	28 ± 10	4 ± 0	0.4

[a]Indicates double fluorescent positive cells (or colonies) for each of the antigens.

[b]Indicates the frequency of double ABC expected by a random distribution of antigen binding receptors among the Ig-bearing population (given by the product of the frequencies of ABC's for each antigen among the Ig-bearing population).

[c]Since the total number of viable bone marrow B-cell colonies in any given experiment was low, pooled data from all of the experiments is given, without standard deviations.

Table 4. Inhibition of antigen binding by anti-immunoglobulin.

Cell Type	Inhibitor Addition	ABC/10^3 KLH	HSF	LDH	NIP-POL	Doubles
BM	None	11.2	8.6	-	-	3.9
	R390 x Ig (1:5)	<0.7	<0.7	-	-	<0.7
	R390 x Ig ABS-MMP*	8.0	8.8	-	-	3.3
Spleen	None	32.2	29.1	-	-	7.6
	R390 x Ig	0.6	<0.6	-	-	<0.6
	R390 x Ig ABS MMP*	20.0	15.1	-	-	3.0
BM	None	-	-	8.0	13.5	2.4
	R390 x MIg	-	-	<0.9	2.6	<0.9
	R390 x Ig + MMP*	-	-	6.6	22.2	0.9
Spleen	None	-	-	14.1	31.2	3.0
	R390 x MIg	-	-	<0.9	0.9	<0.9
	R390 x MIg MMP	-	-	6.8	25.4	0.8
NIP-Specific Cells	None	33.9	-	-	367.6	17.0
	R390 x MIg	1.0	-	-	52.6	<1.0
	R390 x MIg + MMP	23.8	-	-	276.0	11.9

*MMP = mouse myeloma protein pool.

Table 5. Inhibition of antigen binding by unlabelled antigen.

Cell Type	Inhibitor Addition	ABC/10^3 KLH	ABC/10^3 HSF	Doubles
BM	None	8.4	11.7	3.4
	KLH 1st	0.6	7.7	0.6
	KLH 2nd	8.9	7.4	3.2
	HSF 1st	4.5	2.0	0.5
	HSF 2nd	5.0	8.4	2.0
Spleen	None	49.6	23.0	6.2
	KLH 1st	7.2	19.3	1.4
	KLH 2nd	61.8	26.7	7.4
Spleen	None	20	29	4
	KLH 1st	4	42	1
	KLH 2nd	22	31	6
Spleen	None	9	35	2
	KLH 1st	2	32	1
	KLH 2nd	19	26	6
	HSF 1st	6	2	0
	HSF 2nd	8	36	2
Spleen	None	9.3	23	2.5
	KLH 1st	2.3	20.2	<0.8
	KLH 2nd	21.4	17.6	3.8
	HSF 1st	5.8	4.6	<0.8
	HSF 2nd	6.0	39.3	1.7

confirms the earlier observation that the receptors responsible for binding two antigens on double ABC appear to have independent binding sites. The independence of antigen binding receptors was also shown by the lack of strict co-capping (Fig. 1) of the two antigens on double ABCs (Table 7), even though both antigens could be shown to co-cap with immunoglobulin. These results were also confirmed using an analysis of the codistribution (Fig. 2) of two fluorescent ligands on the surface of ABCs (Table 8). As in the case of co-capping, it was found that patches of one fluorescent antigen did not usually correlate with the patches of the other antigen on double ABCs, although ABCs had coincident patches of antigen with fluorescent anti-Ig.

Table 6. Lack of cross-reaction of the receptors on double ABCs among NIP-specific cells.

Exp. No.	Unlabelled Inhibitor Addition	ABC/10^3 KLH	HSF	NIP-POL	Doubles
1	5 mg/ml NIP-POL (1st)	16.1	-	<1	<1
	5 mg/ml NIP-POL (2nd)	28.8	-	217	4
2	none	67	-	369	18
	5 mg/ml NIP-POL (1st)	48	-	<1	<1
	5 mg/ml NIP-POL (2nd)	60	-	385	9
3	none	37	-	274	6
	5 mg/ml NIP-POL (1st)	24	-	<1	<1
	5 mg/ml NIP-POL (2nd)	27	-	249	3
4	none	52	-	319	14
	5 mg/ml NIP-POL (1st)	38	-	4	<1
	5 mg/ml NIP-POL (2nd)	52	-	240	11
	5 mg/ml KLH (1st)	17	-	292	2
	5 mg/ml KLH (2nd)	65	-	269	16
5	none	-	49	314	26
	5 mg/ml NIP-POL (1st)	-	38	3	<1
	5 mg/ml NIP-POL (2nd)	-	38	241	16
	5 mg/ml HSF (1st)	-	15	248	2
	5 mg/ml HSF (2nd)	-	59	316	23

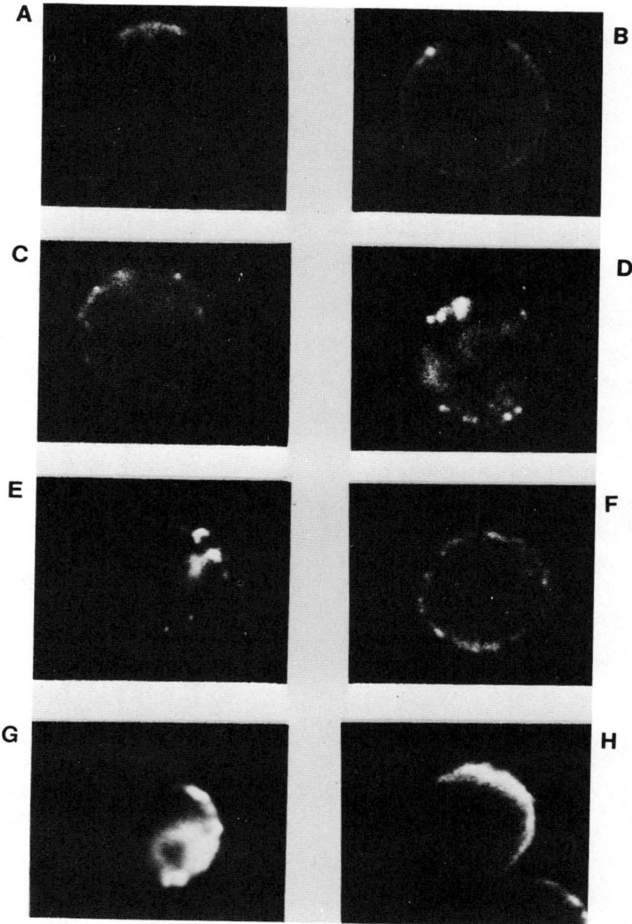

Fig. 1. NIP-specific cells capped for FITC-KLH, cooled, then incubated with TRITC-NIP-POL. Selective excitation for FITC-KLH (b,d,f,h) and selective excitation for TRITC NIP-POL (a,c,e,g) on the same cells showing lack of co-capping of TRITC-NIP-POL by FITC-KLH.

Table 7. Co-capping studies of anti-Ig, Anti-μ, and antigen on ABCs.

Cell Type	1st Ligand	Cap 2nd Ligand	Non-Co-Cap[a]	Co-Cap[a]	% Total Capping of 1st Ligand[b]
Whole Spleen Cells:					
	KLH	anti-Ig	4/11	7/11	73%
	KLH	anti-μ	7/17	10/17	95%
	anti-μ	KLH	8/11	3/11	60%
	HSF	anti-Ig	3/13	10/13	52%
	HSF	anti-μ	5/19	14/19	95%
	anti-μ	HSF	1/17	16/17	74%
	KLH	HSF	2/9	7/9	81%
	HSF	KLH	5/9	4/9	74%
NIP Specific Cells:					
	KLH	NIP-POL	10/12	2/12	65%
	HSF	NIP-POL	16/16	0/16	89%
	NIP-POL	KLH	2/8	6/8	67%
	NIP-POL	HSF	0/9	9/9	44%

a = Only double cells which were capped for the first ligand were scored. Concentrations used were 70 μg/ml NIP-POL and 500 μg/ml of HSF and KLH.

b = i.e., the percentage of 1st ligand cells which had capped that ligand.

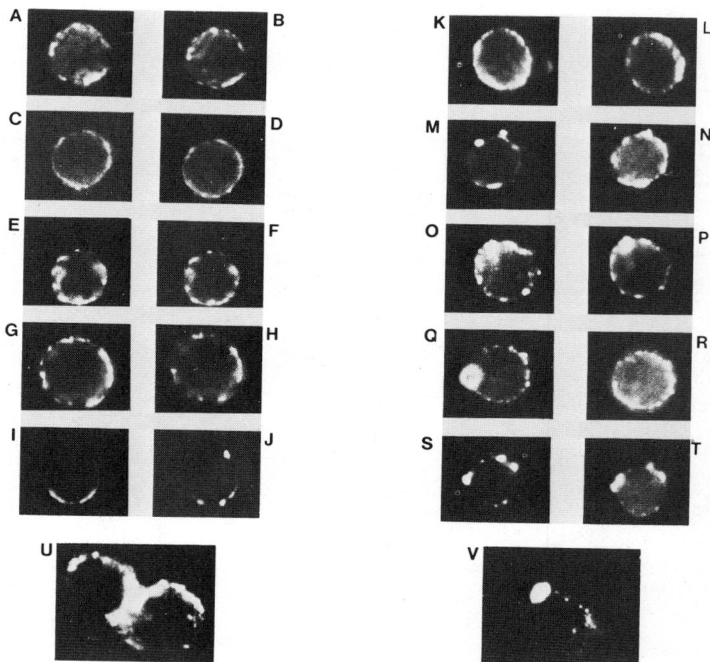

Fig. 2. Codistribution analysis of various ligands and antigen on ABC. (a) Bone marrow control ABCs with TRITC-KLH and (b) FITC-KLH, codistributed; (c) adult spleen control ABCs stained with FITC-HSF and (d) TRITC-HSF, codistributed; (e) neonatal spleen (birth) ABCs stained with TRITC-HSF and (f) FITC-anti-immunoglobulin, codistributed; (g) neonatal spleen (1 day after birth) ABCs stained with FITC-anti-immunoglobulin and (h) TRITC-HSF, codistributed for antigen with extra spots for Ig; (i) NIP-specific adult spleen ABCs stained with TRITC-NIP-POL and (j) FITC-KLH, non-codistributed; (k) NIP-specific adult spleen ABCs stained with TRITC-NIP-POL and (l) FITC-KLH, non codistributed; (m) adult spleen ABCs stained with FITC-LDH and (n) TRITCOHSF, non codistributed; (o) a cytocentrifuged bone marrow derived colony cell stained with TRITC-HSF and (p) FITC-KLH, non-codistributed for a few spots; (q) a neonatal spleen ABC stained with FITC-KLH and (r) TRITC-HSF, non-codistributed; (s) an adult spleen ABC stained with FITC-LDH and (t) TRITC-HSF, non codistributed; (u) two cytocentrifuge bone marrow derived colony cells stained with TRITC-HSF, one of which also binds FITC-KLH, (v) non-codistributed.

Table 8. Co-distribution analysis of antigen and immunoglobulins on antigen binding cells.

Type	Co-distributed	Non-Co-distributed
TRITC-HSF + FITC-HSF or TRITC-KLH + FITC-KLH (Controls)	7	1
TRITC-Antigen + FITC-Anti-Ig	7	1
TRITC-NIP-POL, HSF or KLH + FITC-KLH, HSF or LDH	8	15

IV. BIOSYNTHETIC ORIGIN OF INDEPENDENT RECEPTORS OF MULTIPLE ABCs

The ability of double ABCs to shed and resynthesize their immunoglobulin receptors for both antigens was also tested, and the results indicated that double ABCs required protein synthesis to regenerate their receptors after shedding (Table 9). In these experiments, fluorescent antigens were incubated with the cells, washed and a portion immediately fixed with ρ-formaldehyde (control). The remainder of the cells were cultured overnight in the presence or absence of cycloheximide and checked to show that all surface bound fluorescent material has "capped off". Then, fluorescent reagents were readded to the cells to determine if receptors had been regenerated, particularly on double ABCs. These results are consistent with the notion that independent receptors for both antigens are synthesized by double ABCs, and agrees with the observation described earlier of double ABC-B cell colonies grown in agar. It is unlikely that these double ABC colonies could have picked up receptors from other clones in the agar because of the distances involved in such a transfer.

MATURATION OF MULTIPLE ABCs

Neonatal spleen populations have been studied for antigen-binding capacity in an attempt to determine if the frequency of double antigen-binding cells changes as the population matures (Fig. 3). Just after birth, the frequencies of total ABCs, double ABCs, and Ig-bearing cells were similar to the frequencies seen

Table 9. Shedding and resynthesis of antigen binding receptors.

Organ	Type	Viability	Binding Cells/10^3				
			KLH	HFS	Doubles		
Spleen	Control	99%	31.4	25.4	3.0		
	Shed - 24 hr	57%	1.2	1.2	0		
	Shed Cyclo	30%	0	0	0		
	Readd	-	18.9	11.7	1.3		
	Readd Cyclo	-	1.8	0	0		
			KLH	HSF	Anti-Ig	Doubles (Ag)	Doubles (Ig)
BM	Control (KLH + HSF)	-	12.6	10.0	-	4.2	-
	Shed 18 hr	99%	0.3	0.6	-	0	-
	Shed Cyclo	89%	0.4	0.8	-	0.2	-
	Readd	-	6.3	4.4	-	1.6	-
	Readd Cyclo	-	0.9	0.4	-	0.2	-
	Control (x Ig + HSF)	-	-	18.2	60	-	18.2
	Shed 18 hr	98%	-	0	0	-	0
	Shed Cyclo	19%	-	0.5	1.4	-	0.5
	Readd	-	-	6.2	34	-	6.2
	Readd Cyclo	-	-	0	4	-	0

*Dead cell removal performed before 0.2% eosin exclusion measured.

-Indicates not done

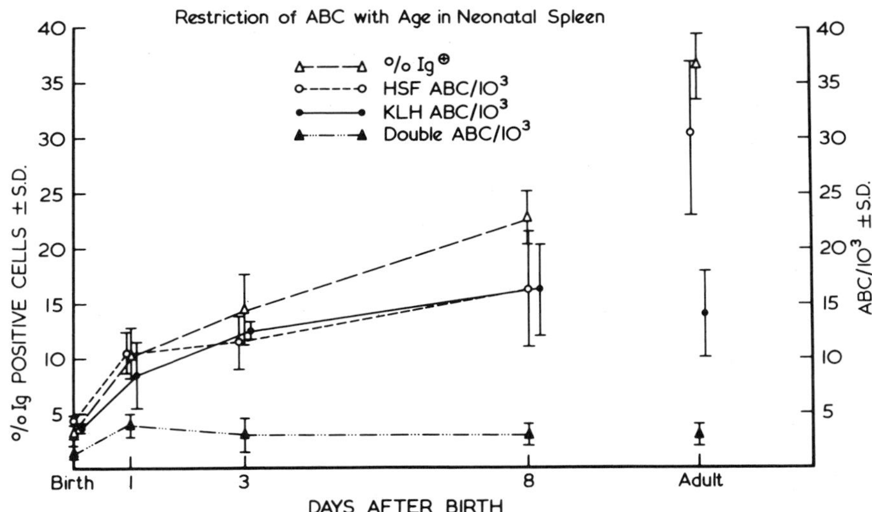

Fig. 3. Restriction of ABCs in neonatal spleen with increasing time after birth. The frequency of Ig-bearing cells, total antigen binding cells and double antigen binding cells at birth is much like that of adult bone marrow. During the first week of life, the frequencies of Ig-bearing cells and total ABCs increase in parallel. However, the frequency of double ABCs, after a slight rise one day after birth, remains relatively constant thereafter. These values are the results form three independent experiments.

in the adult bone marrow. During the first week after birth the proportion of total ABCs and Ig-bearing cells rose sharply so that by 8 days after birth, these cells were about half the adult levels. The proportion of total double ABCs, after a brief rise at day 1, remained constant throughout the test period. Thus, the proportion of total ABCs which are doubles gradually decreases with age (Fig. 4). Tests using pure lymphocyte populations grown in agar colonies from neonatal spleens taken at birth indicate that B lymphocytes are responsible for the antigen binding reaction in that similar frequencies of Ig positive cells and ABCs (including double ABCs) are found.

The maturational state of double ABCs was investigated by 1g sedimentation analysis (DeLuca, unpubl.). Double ABC from 7 day

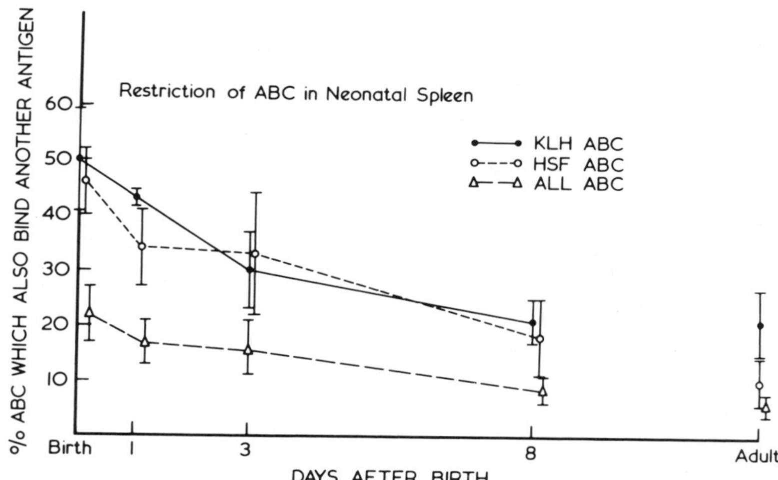

Fig. 4. Restriction of antigen binding capacity of ABCs. The frequency of ABCs which can bind another antigen, i.e., is a double ABC for the antigens shown, decreases progressively during the first week of life, approaching adult levels by 8 days after birth.

germ free spleen sedimented somewhat faster than single ABCs (Fig. 5), and this sedimentation pattern was also found for cells that, on adoptive transfer, must mature in the spleen for 4 days before they become responsive to the thymus independent antigen NIP-POL measured 7 days later ("Pre B cell assay") (Fig. 6). On the other hand, single ABCs sediment at the same rate as cells which are responsive to NIP-POL when it is given with the cells ("B cell assay").

All these data, taken together, support the hypothesis that multiple ABCs are clonally-derived, lymphocyte populations which become more restricted in their antigen-binding capacity as they mature into immunocompetent cells.

V. DISCUSSION

The observations described above can be explained in three ways, each of which is relevant to the molecular basis and functional significance of lymphocytes binding more than one antigen.

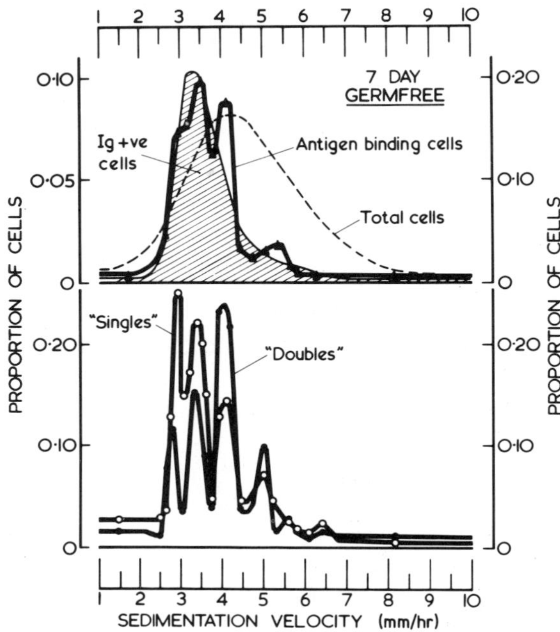

Fig. 5. Sedimentation velocity analysis of spleen cells from 7 day neonatal germ free mice. Upper panel: the sedimentation velocity profiles of Ig-bearing cells, total antigen binding cells, and total cells are shown. Lower panel: the sedimentation velocity profiles of single ABCs (open circles) and double ABCs (closed circles) are given. These profiles represent the means of two independent experiments.

Proposition I. *High Frequency of Antigen-Binding and the Existence of Multiple ABCs is due to Nonspecific Binding*

Many controls have been carried out by investigators performing antigen binding tests to eliminate the possibility of nonspecific antigen binding to lymphoid or nonlymphoid cells. The controls used to eliminate nonlymphoid-cell binding include the use of purified cell populations, e.g., passing the cell population through Sephadex G-10 to remove macrophages (Greenstein et al., 1980). The most direct approach in this area has been the use of pure cloned B cells, macrophages, neutrophils, and eosinophils to determine the cell type responsible for binding (DeLuca, unpubl.). In these experiments, only B cells bound antigen.

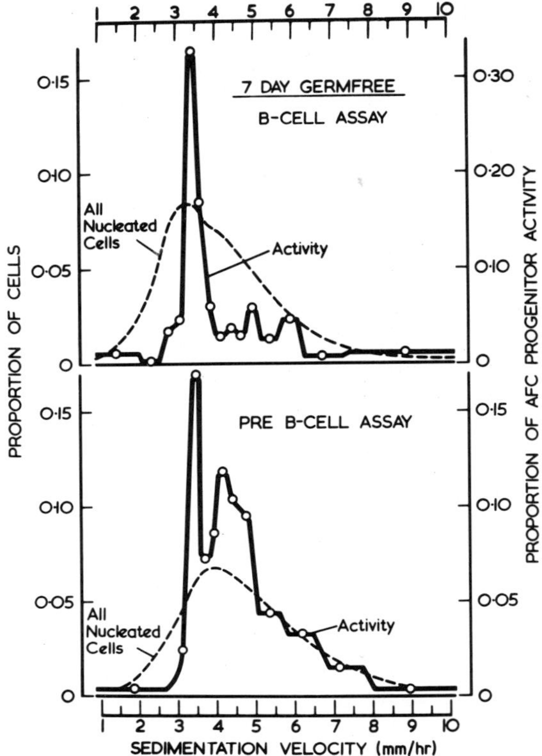

Fig. 6. Sedimentation velocity analysis of the immune response of 7 day neonatal spleen cells to NIP-POL using the B-cell assay (upper panel) and the "Pre-B-cell" assay (lower panel).

In addition, the antigen binding reaction is performed at $4^{o}C$, often in the presence of sodium azide, to prevent active uptake by nonlymphoid cells (Ada, 1970). The identity of an ABC can also be confirmed through the use of specific agents to either stain the ABC in double immunofluroescence, as can be done with anti-immunoglobulin for B cells (Urbain-Vansanten et al., 1974; DeLuca, unpubl.), or by elimination of unwanted populations e.g., T cells with specific anti-T cell antibodies and complement (Kappler and Marrack, 1975).

Once the identity of the cells is established, inhibition of antigen binding by unlabelled antigen and by anti-immunoglobulin reagents has been used in an indirect test to support the notion that an immunoglobulin-like molecule of B cells is responsible for antigen binding. It should be emphasized that some investigators have not been able to inhibit antigen binding with unlabelled antigen (Greenstein et al., 1980), although the vast majority of workers have been able to perform this control. Co-capping (Urbain-Vansanten et al., 1974) and codistribution (DeLuca et al., 1979) of fluorescent antigen with fluorescent anti-immunoglobulin have also provided evidence compatible with the notion that specific immunoglobulin receptors on the B cell surface bind antigen. The possibility that a given cell has acquired its immunoglobulin receptors via passive binding of immunoglobulin (from the external milieu) to F_c receptors has been addressed by appropriate "shedding and resynthesis" experiments (DeLuca, unpubl.) in which ABCs can be shown to actively resynthesize their antigen binding immunoglobulin receptors after having capped and shed their bound antigen/receptor complexes from their surfaces. In addition, since individual clones of ABCs grown in agar can bind antigens in a clonal manner (DeLuca, unpubl.), it seems likely that the cells in the clone must synthesize their own receptors.

However, the tests described above, particularly the inhibition studies, are only indirect in that they are performed at the cell surface. For instance, the ability of glycosylated antigens to bind to the cell may be mediated by nonimmunoglobulin molecules, such as lectins, which have recently been found on lymphocytes (Decker, 1981; Kieda et al., 1978). Unlabelled antigen can, of course, inhibit binding to any receptor, be it immunoglobulin or not. The possibility that anti-immunoglobulin could inhibit antigen binding by steric interference with the ability of the actual antigen receptor to bind the antigen, i.e., by making the receptor inaccessible to the antigen, is another difficulty. A possible example of this phenomenon was reported by McKenzie (1975) in which anti-H2 serum was able to inhibit antigen binding B cells.

A direct approach to the problem of identifying antigen binding receptors in B cells was taken by Marchalonis (1976) in which iodinated nude spleen cells were solubilized and reacted to bound DNP-immunoadsorbent. The material eluted from the DNP immunoadsorbent with DNP-lysine was identified as immunoglobulin through the use of SDS-polyacrylamide gel electrophoresis, as well as affinity chromatography, using antiimmunoglobulin immunoadsorbent to deplete DNP-binding immunoglobulin. This sort of analysis, however, does not address itself to the question of whether or not both receptors for antigen on double-binding cells are immunoglobulin in nature.

Proposition II. *High Frequency of ABCs and Existence of Multiple ABCs is due to Crossreactive Receptors*

It became clear early in the study of antigen binding that the high frequency of ABCs for any given antigen was inconsistent with the idea that a given cell could bind only one antigen. Some investigators (Ada, 1970), taking the lead from functional data gleaned from limiting dilution assays performed in adoptive transfers or in tissue culture, and which indicated a relatively low frequency of responsive B cells, proposed an "antigen concentration gate" above which "irrelevant" antigen binding occurs. The "irrelevant" antigen binding is presumably due to very low affinity immunoglobulin receptors but the binding becomes detectable if enough labelled antigen is added. This interpretation is supported by the well known fact that myeloma proteins can be shown to have a low affinity of binding to a number of different haptens, as well as the observation made by Varga (1973) that a single antibody can bind two haptens at once via independent binding sites on the same molecule.

The postulates outlined above make certain predictions about the ability for increasing concentration of antigen to yield an increasing number of antigen binding cells. If, indeed, antigen is bound nonspecifically or only by low affinity receptors at high concentrations, the frequency of ABCs would continue to rise as a function of antigen concentration. If, on the other hand, the binding of antigen is mediated by saturable specific receptors, the frequency of ABCs will eventually plateau such that increasing the concentration of antigen no longer increases the frequency of ABCs. Several reports (Ada, 1970; DeLuca, unpubl., Deluca et al., 1979), including analysis of antigen binding with a fluorescence activated cell sorter (Greenstein et al., 1980), indicate that even at high antigen concentrations, the frequency of ABCs reaches a plateau value. Some workers using rather more insensitive assays such as ^{3}H autoradiography (Diener and Paetkau, 1972), find a plateau frequency somewhat lower than others. This may have to do with different sensitivities of various assays. It should also be pointed out that the frequency of ABCs found, even under saturation conditions, will depend on the size of the antigen and the number of epitopes it carries, i.e., one would expect that a large protein antigen would bind to many more cells than a small hapten. This, indeed appears to be the case, since, for example, the binding of TRITC NIP-polymerized flagellin (NIP-POL) can be reduced by 90% by preincubation with unlabelled POL (DeLuca, unpubl.).

The question of crossreactive receptors has been addressed directly on cell surfaces in two different ways. The ability of a high excess of antigen (NIP-POL) to inhibit the binding of key hole limpet hemocyanin (KLH) or horse spleen ferritin (HSF) to purified NIP-specific ABC was tested by DeLuca (unpubl.). It was found that while the binding of fluorescent labelled NIP-POL to the double ABC was completely inhibited by unlabelled NIP-POL, the ability of the cells to bind the other antigen was not changed. In a similar fashion, the binding of KLH or HSF could be inhibited without affecting the ability of double ABCs to bind NIP-POL. In addition, it was shown by DeLuca (unpubl.) and Urbain-Vansanten et al., (1974) that the antigens bound to double ABCs could be induced to cap independently of one another, although the binding of either antigen on all ABCs was coincidental with immunoglobulin as shown in double immunofluorescence.

The above experiments may be interpreted to indicate that antigen binding to multiple ABCs occurs via two independent binding sites, but, as pointed out earlier, this binding is not necessarily due to immunoglobulin, nor is it proven that more than one molecular species of receptor actually exists on the surface of these cells.

Proposition III. *High frequency of ABCs and Existence of Multiple ABCs is due to Multiple Immunoglobulin Receptors Expressed by Immature B Lymphocytes*

The most powerful evidence gathered so far to support the idea that a given ABC can bind multiple antigens via independent receptors comes from the cross-inhibition experiments of DeLuca (unpubl.) described above, and the cocapping and codistribution data of DeLuca (unpubl.) and Urbain-Vansanten (1974).

Since the frequency of ABCs, taken as a percentage of the total immunoglobulin bearing population, is highest among bone marrow B cells and neonatal (at birth) B cells as compared to spleen B cells, one might expect that the frequency of double ABCs would be higher among the immature cells, presuming a random distribution of antigen binding capacity. This, indeed, is the case. A study made of maturing B lymphocytes in the spleen during the first week of life (DeLuca, unpubl.) indicates that the ability of the cells to bind antigens becomes progressively more restricted as the B cell population matures. Thus, at birth the frequencies of immunoglobulin-bearing cells, total ABCs, are much like those of adult bone marrow. However, during the first week of life, the frequencies of immunoglobulin bearing cells and total ABCs increase in a parallel manner, while the frequency of double ABCs remains the same. The overall effect of this turn of events is that the average B cell in the mature spleen is more restricted in the number of antigens that it can recognize than the average bone marrow B cell.

With this idea in mind, the sedimentation velocity profiles of single and double ABCs from spleens of 7 day germ free mice were determined by DeLuca (unpubl.) along with the profiles of cells responsive to the thymus independent antigen, NIP-POL in irradiated recipients. It was found that single ABCs co-sedimented with cells which were immediately responsive to the antigen if it was given at the time of cell transfer. However, double ABCs did not co-sediment with these cells, but they did cosediment with cells that required 4 days between their injection into the irradiated recipients and the administration of antigen before they could respond.

The results described above and in the previous sections fit most easily within a framework of continual restriction of the multiplicity of antigen recognizing units on B cells during maturation. This concept is consistent with the results of Couderc et al. (1975, 1977, 1979). These investigators found that early in the immune response to two non-crossreacting haptens, a small subpopulation of plaque-forming cells (PFCs) were capable of producing lytic antibodies to two types of easily distinguishable red blood cells, each of which was coupled to one of the two haptens used in the immunization. The double lysis of the indicator cells was apparently mediated by independent antibodies as shown in cross inhibition tests similar to those done for antigen binding (Couderc et al., 1975, 1979). As in the antigen binding studies described above, the frequency of double PFCs found when recipients reconstituted with bone marrow cells were challenged with antigen was much higher than if spleen cells were used to reconstitute the recipients. Initially, as many as 30% of the total PFCs generated from recipients of bone marrow cells after challenge with two antigens are double PFCs, but with continued maturation of the cells, the frequency of double PFCs continually fell until a level approaching that found for normal spleen cells was reached (Couderc et al., 1977).

Restriction of antigen recognition capacity by maturing B cells, at least in the case of antigen binding cells in neonatal mice, appears to be antigen independent since the animals had not been previously exposed to the antigens used. Since it has been shown that during B cell maturation new clonotypes arise which are absent from the neonatal repertoire (Cancro et al., 1979), it may be that diversification of the repertoire is occurring during B cell maturation. The idea of a genetically patterned acquisition of antigen recognizing capacity is supported by the fact that the ability of a given mouse strain to bind a panel of antigens is a stable property of that strain (Cohen et al., 1977).

However, a considerable body of evidence indicates that the presence of antigen can modify the specificity of cells which are mature enough to respond to antigens. Some investigators have reported that antigens can select for the propagation of variant clones which are specific for the selecting antigens (Cunningham and Fordham, 1974; Cunningham, 1974, 1976a,b). This phenomenon has also been reported in the context of selection of individual single specificity PFCs from parent double specificity PFCs (Couderc et al., 1979). Indeed, some workers have shown that immunoglobulin producing cells found very early in the response do not have specificity for the immunizing antigen, but these cells appear to give rise to antigen specific cells later in the response (Miller et al., 1975; Bernard et al., 1979).

An important question in all of these studies is how the initial clones of cells become triggered. It seems quite possible that the initial clones are triggered by nonspecific factors, such as T cell factors, or by mitogen specific (rather than antigen specific) receptors which act to amplify the B cell response in a "polyclonal" manner. Antigen may then act to "focus" or select those clones specific for the antigen via the affinity of their immunoglobulin receptors by the antigen (Coutinho and Moller, 1974; Cammisuli and Henry, 1978). It is interesting to note in this context the results of Forni (1979) and Coutinho et al. (1978, 1979), who found a very high percentage (10-15%) of anti-dextran idiotype binding spleen B cells, as well as double anti-idiotype binding fetal liver and spleen cells for several idiotypes. Anti-idiotype binding results in activation of a large percentage of the binding cells, but only a few of the plasma cells resulting from the activation secreted idiotype bearing Ig. Since anti-idiotype binding occurs on some immature Ig negative lymphoid cells, it has been concluded that the determinants recognized by the anti-idiotypes are not an immunoglobulin, but are on "growth receptors" bearing determinants crossreactive with germ line V genes. It was postulated that interactions of these receptors with anti-idiotype networks may regulate B growth. It seems to us that the failure of these workers to detect immunoglobulin on some of their idiotype bearing cells by immunofluorescence may have to do with incomplete expression of Ig through the plasma membrane of immature B cells such that Ig isotype class-specific determinants were not available for reagents specific for that part of the molecule. In any case, these results imply that multiple idiotypes are expressed in immature populations of lymphoid cells, and that these may be responsible for our antigen binding results.

Although studies involving both antigen binding and assay of secreted antibodies by single cells imply the presence of multiple receptors, no study has been done so far to show directly (1) how many different immunoglobulin molecules specific for different antigens can be made by a single cell during its life-time, and (2) if these molecules can be shown to exist, if they are capable of triggering the cell. If it can be shown that several functional receptors specific for different antigens can be made by a single cell, the current concepts involving the genetics of V-region diversity during B cell maturation may be extended to include a longer period of B cell maturation. It is now widely accepted that the formation of an active immunoglobulin gene in myeloma cells involves the somatic recombination of V-region genes and C-region genes (Capra and Kindt, 1975; Bernard et al., 1978; Hozumi and Tonegawa, 1976). This rearrangement appears to occur sometime between the beginning of lymphoid development and the end of the B cell lineage, the plasma cell. It has been proposed that the switch in C region class during clonal expansion (Gearhart et al., 1975; Wahl et al., 1978) is due to translocation of the V-region from a μ chain gene to a γ chain gene with deletion of the μ gene (Coleclough et al., 1980; Maki et al., 1980; Davis et al., 1980; Cory and Adams, 1980; Rabbitts et al., 1980; Kataoka et al., 1980). These ideas would provide a ready explanation for allelic exclusion, immunoglobulin class "switchover" and some aspects of the generation of diversity, i.e., through D and J region coding (Rao et al., 1979; Joho et al., 1980).

However, rather little has been done to investigate the genetic events which are involved during that crucial late fetal and early neonatal period when the immune system must expand both its V-region repertoire (Press and Klinman, 1974; Sigal, 1977) and its total population of immunoglobulin receptor bearing B cells. It has been pointed out that this process must occur during a relatively small number of cell divisions (Cohen et al., 1977). If this is true, then it may be that the insertion of "mini-genes" coding for the hyper-variable regions (Kabat et al., 1978, 1979; Wu et al., 1979), recombination between V regions during cell division (Cook and Scharff, 1977; Seidman et al., 1978; Max et al., 1979) translocation of V-region genes to C-region genes (Capra and Kindt, 1975; Hozumi and Tonegawa, 1976; Bernard et al., 1978) or selective mRNA "processing" (Seidman and Leder, 1980; Choi et al., 1980) may occur many times during the early life of B cells. Such switching may occur via an episomal mechanism whereby a loop of DNA containing many V regions may sequentially bring V regions into register with the transcription C region through reversible ligature with the J region, allowing transcription of the complete message. The order of V regions, V_1, V_2, etc, may account for the ordered acquisition of the B cell repertoire during development (Cooper et al., 1979). If this switching of V region genes occurs continuously during clonal expansion during B cell development, multiple stable Ig mRNA transcripts may result. The products of these may be expressed for a time on the cell surface, in much the same way as μ chain is expressed for a time along with γ chain during Ig class switchover (Wahl et al., 1978). Interestingly, fragments of several different light chains can be found in some myelomas (Alt et al., 1980). An interesting idea would involve a variation of a postulate made by Burrows et al. (1979) in which the sequential expression of V_H regions of surface Ig negative pre-B cells is followed by a selection of a number of randomly chosen V_L regions on the light chains which these cells subsequentially synthesize. It may be that the cell then expresses many V_H-V_L combinations until it "fixes" the proper combination during clonal expansion and differentiation. Antigen may provide some driving force in the selection of which V_H-V_L combination is chosen by continually stimulating the propagation of those cells making the V_H-V_L with the strongest affinity for the antigen.

VII. REFERENCES

Ada, G. L. (1970). Antigen binding cells in tolerance and immunity. *Transplant Rev.*, 5, 105-129.

Alt, F. W., Enea, V., Bothwell, A. L. M., and Baltimore, D. Activity of multiple light chain genes in murine myeloma cells producing a single, functional light chain. *Cell*, 21, 1-12.

Bernard, O., Hozumi, N., and Tonegawa, S. (1978). Sequences of mouse immunoglobulin light chain genes before and after somatic changes. *Cell*, 15, 1133-1144.

Bernard, J., Jeannesson, P., Thierness, N., Zagury, D., Ternynck, T. and Avrameas, S. (1979). Subpopulations of Ig-secreting cells induced by peroxidase immunization: discrimination according to antibody storage and secretion. *Immunol.*, 36, 719-727.

Burnet, F. M. (1959). "The Clonal Selection Theory of Acquired Immunity". *Cambridge University Press*, Cambridge, England.

Burrows, P., Lejeune, M., and Kearney, J. F. (1979). Evidence that murine pre-B cells synthesize heavy chains but nog light chains. *Nature*, 280, 838-841.

Byrt, P. and Ada, G. L. (1969). An *in vitro* reaction between labelled flagellin or haemocyanin and lymphocyte-like cells from normal animals. *Immunology*, 17, 503-516.

Cammisuli, S. and Henry, C. (1978). Role of membrane receptors in the induction of an *in vitro* antihapten response II. Antigen-immunoglobulin receptor interaction is not required for B memory cell proliferation. *Eur. J. Immunol.*, 8 662-666.

Cancro, M. P., Wylie, D. E., Gerhard, W., and Klinman, N. R. (1979). Patterned acquisition of the antibody repertoire: Diversity of the hemagglutin specific B cell repertoire in neonatal Balb/c mice. *Proc. Natl. Acad. Sci. USA*, 76, 6577-6581.

Capra, J. D. and Kindt, T. S. (1975). Antibody diversity: can more than one gene encode each variable region? *Immunogenetics*, 1, 417-427.

Cohen, J. E., E'Eustachi, P., and Edelman, G. (1977). The specific antigen-binding cell populations of individual mouse spleens: repertoire composition, size and genetic control. *J. Exp. Med.*, 146, 394-411.

Coleclough, C., Cooper, D., and Perry, R. P. (1980). Rearrangement of immunoglobulin heavy chain genes during B-lymphocyte development as revealed by studies of mouse plasmacytoma cells. *Proc. Natl. Acad. Sci. USA*, 77, 1422-1426.

Cook, W. D. and Scharff, M. D. (1977). Antigen-binding mutants of mouse myeloma cells. *Proc. Natl. Acad. Sci. USA*, 74, 5687-5691.

Cooper, M. G., Ada, G. L., and Langman, R. E. (1972). The incidence of hemocyanin-binding cells in hemocyanin tolerant rats. *Cell Immunol.*, 4, 239-303.

Cory, S. and Adams, J. (1980). Deletions are associated with somatic rearrangment of immunoglobulin heavy chain genes. *Cell*, 19, 37-48.

Couderc, J., Bleux, C., Birrien, J. L., and Liacopoulos, P. (1975a). The potentiality of antibody forming cells I. Bispecific cell occurrence in double stimulated cultures of syngeneic or allogeneic spleen cells of the mouse. *Immunol.*, 29, 653-664.

Couderc, J., Bleux, C., and Liacopoulos, P. (1975b). The potentiality of antibody-producing cells II. Evidence for two antibody molecules of different specificities secreted by micromanipulated bispecific mouse spleen cells. *Immunol.*, 29, 665-674.

Couderc, J., Birrien, J. L., Bleux, C., and Liacopoulos, P. (1977). Development of responsiveness and incidence of bi-specific cells as revealed by *in vitro* assessment of the maturation of mouse bone marrow cells. *Cell Immunol.*, 28, 248-257.

Couderc, J., Bleux, C., Ventura, M., and Liacopoulos, P. (1979). Single mouse cells producing two antibody molecules and giving rise to antigen driven intraclonal variation after immunization with two unrelated antigens. *J. Immunol.*, 123, 173-181.

Coutintio, A. and Moller, G. (1974). Immune activation of B cells: evidence for "one nonspecific triggering signal" not delivered by the Ig receptors. *Scand. J. Immunol.*, 3, 122-146.

Coutinho, A., Forni, A., and Blomberg, B. (1978). Shared antigenic determinants by mitogen receptors and antibody molecules to the same thymus dependent antigen. *J. Exp. Med.*, 148, 862-870.

Cutinho, A. and Forni, L. (1979). Basel Institute for Immunology Annual Report, Vol. 88.

Cunningham, A. J. (1974). The generation of antibody diversity: its dependence on antigenic stimulation. *Contemp. Topics Immunol.*, 3, 1.

Cunningham, A. J. (1976a). Evolution in microcosm: the rapid somatic diversification of lymphocytes. *Cold Spring Harbor Symp. Quant. Bio.*, 41, 761-770.

Cunningham, A. J. (1976a). In "Generation of Antibody Diversity: A New Look", Cunningham, A. J. (Ed.). Academic Press, London. p. 89.

Cunningham, A. J. and S. A. Fordham. (1974). Antibody cell daughters can produce antibody of different specificities. *Nature*, 250, 669-671.

Davis, M. M., Calame, K., Early, P. W., Kivant, D. L., Joho, R., Weissman, I. L., and Hood, L. (1980). An immunoglobulin heavy-chain gene is formed by at least two recombinational events. *Nature*, 283, 733-739.

Decker, J. M. (1981). Lectin-like molecules specific for the oligosaccharide moiety of the fetal α-globulin fetuin. *Molecular Immunol.*, 17, 803-808.

DeLuca, D., Miller, A., and Sercarz, E. E. (1975). Antigen binding to lymphoid cells from unimmunized mice IV. Shedding and reappearance of multiple antigen binding Ig receptors of T and B lymphocytes. *Cell Immunol.*, 18, 186-303.

DeLuca, D., Warr, G. W., and Marchalonis, J. J. (1979). The immunoglobulin-like T-cell receptor II. Codistribution of Fab determinants and antigen on the surface of antigen binding lymphocytes of mouse thymus. *J. Immunogen.*, 6, 359-372.

Diener, E. and Paetkau, V. H. (1972). Antigen recognition II. Early surface receptor phenomena induced by binding of a tritium labelled antigen. *Proc. Nat. Acad. Sci. USA*, 69, 2364-2368.

Donald, D., King, D. J., and Beck, J. S. (1974). Antigen-binding small lymphocytes in the guinea pig II. The immunological response to purified protein derivative of mammalian tuberculin *Immunol.*, 27, 87-97.

Dwyer, J. M. and Mackay, I. R. (1972). Validation of autoradiography for recognition of antigen-binding lymphocytes in blood and lymphoid tissues. *Clin. Exp. Immunol.*, 10, 581-597.

Forni, L., Cazenave, P-A., Cosenza, H., Forsbeck, K., and Coutinho, A. (1979). Expression of V region-like determinants on Ig-negative precursors in murine fetal liver and bone marrow. *Nature*, 280, 241-245.

Gearhart, P. J., Sigal, N. H., and Klinman, N. R. (1975). Production of antibodies of identical idiotype but diverse immunoglobulin classes by cells derived from a single stimulated B-cell. *Proc. Natl. Acad. Sci. USA*, 72, 1707-1711.

Greenstein, J. L., Leary, J., Horan, P., Kappler, J. W., and Marrack, P. (1980). Flow sorting of antigen binding B cell subsets. *J. Immunol.*, 124, 1472-1481.

Hozumi, N. and Tonegawa, S. (1976). Evidence for somatic rearrangement of immunoglobulin genes coding for variable and constant regions. *Proc. Nat. Acad. Sci. USA*, 73, 3628-3632.

Joho, R., Weissman, I. L., Early, P., Cole J., and Hood, L. (1980). Organization of K light chain genes in germ line and somatic tissue. *Proc. Nat. Acad. Sci. USA*, 77, 1106-1110.

Julius, M. H. and Herzenberg, L. A. (1974). Isolation of antigen-binding cells from unprimed mice. *J. Exp. Med.*, 140, 904-920.

Julius, M. H., Janeway, Jr., C. A., and Herzenberg, L. A. (1976). Isolation of antigen-binding cells from unprimed mice II. Evidence for monospecificity of antigen-binding cells. *Eur. J. Immunol.*, 6, 288-292.

Kabat, E. A., Wu, T. T., and Bilofsky, H. (1978). Variable region genes for the immunoglobulin framework are assembled from small segments of DNA--A hypothesis. *Proc. Nat. Acad. Sci. USA*, 75, 2429-2433.

Kabat, E. A., Wu, T. T., and Bilofsky, H. (1979). Evidence supporting somatic assembly of the DNA segments (mini genes) coding for the framework and complementarity-determining segments of immunoglobulin variable regions. *J. Exp. Med.*, 149, 1299-1313.

Kappler, J. W. and Marrack (Hunter), P. C. (1975). Functional heterogeneity among the T derived lymphocytes of the mouse III. Helper and suppressor T cells activated by concanavalin A. *Cell Immunol.*, 18, 9-20.

Kataoka, T., Kawakami, T, Takahashi, N., and Honjo, T. (1980). Rearrangment of immunoglobulin μ-chain gene and mechanism for heavy-chain class switch. *Proc. Nat. Acad. Sci. USA*, 77, 919-923.

Kennedy, J. C., Till, J. E., Siminovitch, L., and McCulloch, E. A. (1966). The proliferative capacity of antigen-sensitive precursors of hemolytic plaque forming cells. *J. Immunol.*, 96, 973-980.

Kieda, C. M. T., Bowles, D. J., Rivd, A., and Sharon, N. (1978). Lectins in lymphocyte membranes. *FEBS Letters*, 94, 391-396.

Lawrence, D. A., Speigelberg, H., and Weigle, W. O. (1973). 2,4-dinitrophenol receptors on mouse thymus and spleen cells. *J. Exp. Med.*, 137, 470-482.

Maki, R., Traunecker, A., Sakano, H., Roeder, W., and Tonegawa, S. (1980). Exon shuffling generates an immunoglobulin heavy chain gene. *Proc. Nat. Acad. Sci. USA*, 77, 2138-2142.

Marchalonis, J. J. (1976). Isolated radio-iodinated surface immunoglobulins of murine bone-marrow derived lymphocytes which bind the 2,4-dinitrophenyl hapten. *Immunochemistry*, 13, 667-670.

Max, E. E., Seidman, J. G., and Leder, P. (1979). Sequences of five potential recombination sites encoded close to an immunoglobulin K constant region gene. *Proc. Nat. Acad. Sci. USA*, 76, 3450-3454.

McKenzie, I. F. C. (1975). Ly 4.2: A cell membrane alloantigen of murine B lymphocytes. II. Functional studies. *J. Immunol.*, 114, 856-863.

Miller, H. P., Ternynck, T., and Avrameas, S. (1975). Synthesis of antibody and immunoglobulins without detectable antibody function in cells responding to horseradish peroxidase. *J. Immunol.*, 114, 626-634.

Naor, D. and Sulitzeanu, D. (1967). Binding of radioiodinated bovine serum albumin to mouse spleen cells. *Nature*, 214, 687-688.

Playfair, J. H. L., Papermaster, B. W., and Cole, L. J. (1965). Focal antibody production by transferred spleen cells in irradiated mice. *Science*, 149, 998-1000.

Press, J. L. and Klinman, N. R. (1974). Frequency of hapten-specific B-cells in neonatal and adult mouse spleens. *Eur. J. Immunol.*, 4, 155-159.

Seidman, J. G., Leder, A., Nau, M., Norman, B., and Leder, P. (1978). Antibody diversity. The structure of cloned immunoglobulin genes suggests a mechanism for generating new sequences. *Science*, 202, 11-17.

Sigal, N. H. (1977). The frequency of p-azophenylarsonate and dimethylamino-naphalene-sulfonyl-specific B-cells in neonatal and adult Balb/c mice. *J. Immunol.*, 119, 1129-1133.

Rabbitts, T. H., Forster, A., Dunnick, W., and Bentley, D. L. (1980). The role of gene deletion in the immunoglobulin heavy chain switch. *Nature*, 283, 351-356.

Rao, D. N., Rudikoff, S., Krutzsch, H., and Potter, M. (1979). Structural evidence for independent joining region gene in immunoglobulin heavy chains from anti-galactan myeloma proteins and its potential role in generating diversity in complementary-determining regions. *Proc. Nat. Acad. Sci. USA*, 76, 2890-2894.

Rutishauser, U., Millette, C. F., and Edelman, G. M. (1972). Specific fractionation of immune cell populations. *Proc. Nat. Acad. Sci. USA*, 69, 1596-1600.

Shearer, G. M., Cudkowicz, G., and Priore, R. L. (1969). Cellular differentiation of the immune system of mice II. Frequency of unipotent splenic antigen sensitive units after immunization with sheep erythrocytes. *J. Exp. Med.*, 129, 185-199.

Urbain-Vansanten, G., Richard, C., Bruyno, C., Hoogen, V., van Acker, A., and Urbain, J. (1974). High number of antigen-binding cells in unimmunized mice and possible occurrence of multispecific lymphocytes. *Ann. Immunol. (Inst. Pasteur)*, 125C, 885-893.

Varga, J. M., Konigsberg, W. H., and Richards, F. C. (1973). Antibodies with multiple binding functions: Induction of single immunoglobulin species with structurally dissimilar haptens. *Proc. Nat. Acad. Sci. USA*, 70, 3269-3263.

Wahl, M. R., Forni, L., and Loor, F. (1978). Switch in immunoglobulin class production observed in single clones of committed lymphocytes. *Science*, 199, 1078-1082.

Wu, T. T., Kabat, E. A., and Bilofsky, H. (1979). Some sequence similarities among cloned mouse DNA segments that code for λ and K light chains of immunoglobulins. *Proc. Nat. Acad. Sci. USA*, 76, 4617-4621.

EXPRESSION OF ANTIBODY DIVERSITY BY SINGLE B CELLS[1]

P. Liacopoulos, C. Bleux, M. Ventura, C. Desaymard, and J. Couderc

Institut d'Immuno-Biologie
(U. 20 I.N.S.E.R.M., L.A. 143 C.N.R.S.)
Hôspital Broussais
96 Rue Didot
75674 Paris Cedex 14, France

[1]This work was supported by Grant ATP No. 3615 from C.N.R.S. and Grant C.R.L. No. 77-5-70 from I.N.S.E.R.M.

I. INTRODUCTION

In order to explain the fundamental biological problem of generation of antibody diversity, or more exactly of antibody specificity, two theories have been advocated for many years. The germ line theory puts forward that all diversity is encoded in the V region genes, whereas the somatic mutation theory states that diversity is originated from mutations occuring on a limited number of V region genes during ontogeny. Discovery in recent years that genome contains a large number of V genes as well as multiple D and J minigenes, a combinational association of germ line elements has prevailed as the main mechanism of the generation of diversity, without ruling out the possibility of some contribution of somatic mutations, since such mutations have also been observed in a significant frequency in germ line genes (Schilling et al., 1980).

Both theories, however, agreed that only one pair of H and L, V genes and C genes were active after rearrangement and expressed in each particular B cell. The choice of this rearrangement might occur at the differentiation stage of pre-B cell and, thereafter, each of these cells would give rise to a particular clone expressing the corresponding antibody specificity or, if a mutation occurs on the V gene, the clone would then express only the new specificity.

Our studies were focused on whether one particular B cell can only express a single pair of V genes. We were interested in this question because the above assumptions were based on what is observed during the phase of mature antibody response when each B cell produces only one type of antibody molecule even after double immunization. It remained possible, therefore, that examination of cells during the early period after immunization with two antigens would detect some cells which could simultaneously respond to both antigens. In such a case restriction to the production of one antibody per cell would be the result of differentiation of B cells toward mature plasma cells rather than from permanent rearrangement of one single pair of H and L V genes at the stage of the pre-B cell.

II. PRODUCTION OF TWO ANTIBODY MOLECULES BY SINGLE CELLS

Immunization of mice with two unrelated antigens can indeed result in the regular appearance of spleen cells simultaneously reacting to both antigens. For appearance of such cells it is necessary that the two antigens induce a rapid and vigorous cell response. This can be achieved by using erythrocytes for eliciting primary response (Couderc et al., 1975) or hapten-carrier conjugates in solution for challenging animals previously primed with the same conjugates in mixture with complete Freund's adjuvant (Couderc et al., 1979). In both cases, these double cells appear temporarily during the 3rd and 5th day after the eliciting injection; the cellular response being studied with the hemolytic plaque formation method, using as indicator native or hapten-substituted red cells. The frequency of PFC always remains low, not exceeding 0.5-2% out of the total PFC.

Use of hapten-carrier conjugates as immunogens permitted a satisfactory characterization of the specificity of the cellular response. The four haptens used, trinitrophenyl (TNP) sulfonic acid (Sulf), aminophenyl-βD-lactoside (Lac), and phosphorylcholine (PC), structurally very dissimilar, were usually conjugated to KLH for immunization. In all these experiments PFCs were developed by the indirect method using a rabbit anti-mouse IgG enhancing serum. It can be seen on Table 1 that, in each case, a significant number of double cells were detected which simultaneously lysed both types of indicator erythrocytes (sheep and pigeon RBC conjugated with the corresponding haptens). This double lysis does not stem from the use of the same carrier used for immunization (KLH) since, as shown in the last row of Table 1, when a mixture of Lac-KLH and Sulf-RGG (rabbit gamma globulin) was used for immunization of mice, the same frequency of double PFC was again found. Furthermore the carriers were different for each phase of the study: KLH or RGG for immunization, SRBC or PRBC for detection of PFC, and bovine serumalbumin (BSA) or lysozyme (LZ) for specific inhibition. Absence of cross-reactions between the haptens was further sought for by specific inhibition experiments at the single cell level. Double PFCs were micromanipulated from the initial revealing medium into a second medium containing the indicators plus one soluble inhibitor (TNP-BSA or Sulf-LZ) and then transferred into a third medium containing the indicators alone (Table 2). Almost all double PFCs which remained active in the three successive media (50-60% of micromanipulated cells), lysed only one indicator when transferred into the second medium containing the alternate soluble inhibitor and again lysed both indicators after transfer into the third medium which did not contain inhibitor. Therefore, the inhibition of antibody secreted by double PFC was not competitive, since the inhibition of one lytic activity did not interfere with the lysis of the other indicator. This ruled out the possibility that antibody secreted by double PFCs was a polyfunctional antibody in which inhibition with hapten-carrier conjugate competively inhibits both reactivities (Varga et al., 1973). Absence of competitive inhibition in double PFCs clearly indicated production of two different antibody molecules.

Further evidence as to the production of two different antibody molecules by double cells was obtained from experiments where double PFCs from double immunized animals were individually

Table 1. Number and specificity of indirect PFC/10^6 spleen cells detected at the peak of the response (4th to 5th day) after secondary challenge of the mice immunized with one of the antigen pairs. In each experiment 4 to 8 x 10^5 spleen cells from 3 mice were examined.

Immunizing Antigens	N° of Experiments	SPECIFICITY OF PFC					
		Anti-TNP	Anti-Sulf	Anti-Lac	Anti-Pc	Doubles	% of Doubles Out of Total PFC
TNP-KLH + Sulf-KLH	11	551±175[x]	954±258	-	-	8.5±3.4	0.37±0.12
TNP-KLH + Lac-KLH	11	166± 48	-	332±135	-	2.0±0.8	0.40±0.15
Sulf-KLH + Lac-KLH	4	-	760±534	439±371	-	15.4±9.9	1.34±0.52
TNP-KLH + PC-KLH	4	242±113	-	-	306±220	7.0±3	1.2 ±0.83
Sulf-RGG + Lac-KLH	6	-	159± 71	137± 58	-	3.75±1.4	1.2 ±0.9

(x) average (± S.E.)

Table 2. Lytic activity of individual double PFC during transfer from the initial revealing medium containing both indicators (TNP-PRBC + Sulf-SRBC) to the second medium containing the same indicators + one soluble inhibitor (either TNP-BSA, top row or Sulf-LZ, bottom row) and then to the third medium containing both indicators alone.

Initial medium TNP-PRBC + Sulf-SRBC N° of double PFC	Second medium TNP-PRBC + Sulf-SRBC + TNP-BSA				Third medium TNP-PRBC + Sulf-SRBC			
	RBC lysis				RBC lysis			
	Both	TNP-PRBC	Sulf-PRBC	None	Both	TNP-PRBC	Sulf-SRBC	None
68	0	0	56	12	35	6	8	19
100%	0	0	82	18	53	8	12	27
	+ Sulf-LZ							
82	2	67	0	13	47	11	6	18
100%	3	81	0	16	57	13	8	22

cultured. After secondary immunization of mice with TNP-KLH and PC-KLH, individual PFCs of the three types observed (anti-TNP, anti-PC, and doubles) were cultured in the presence of both antigens according to the method previously described (Couderc et al., 1979). In these experiments, only a proportion of 20-25% of individually cultured PFCs yield a progeny, the others, for unknown reasons, did not divide and disappear. Those double PFCs which divided mainly generated daughter PFCs producing antibody of one or other specificity, thus showing the temporary nature of the double production (Table 3). The results observed after microculture of double anti-TNP and anti-SRBC PFCs appearing after primary immunization of mice with TNP-SRBC (Liacopoulos et al., 1976) are presented in the same table. Consequently, of the double lytic activity of parental PFCs, only one was inherited in the daughter PFCs after division. This also rules out the possibility of production of polyfunctional antibody by parental PFCs since the latter lysed both indicators whereas antibody produced by daughter PFCs lyzed only one indicator.

The production of two antibody molecules by an individual PFC gives rise to the problem of the molecular mechanism of this double production. Indeed, the well known phenomenon of allelic exclusion indicates that only one chromosome is active in antibody producing cells and this concept has been supported by work disclosing that in myeloma cells only one chromosome is rearranged (Brack et al., 1978). However, more recent studies showed that in some myelomas, lymphomas and especially in an important proportion of normal spleen cells, both chromosomes are rearranged (Hurwitz et al., 1980), although it seems that rearrangement of one of the chromosomes would not be functional (Van Ness et al., 1981). Therefore, we started to examine whether in double PFCs both chromosomes are expressed. Hybrid (C57BlxDBA/2)F_1 mice were secondarily immunized with TNP-KLH and PC-KLH and PFCs were enhanced either with mixtures of anti-mouse γ and anti-mouse μ, with anti-μ and specific anti-DBA/2 allotype (Ig-1^c) or anti-μ and specific anti-C57Bl (Ig-1^b) sera. As shown in Table 4, cell populations, enhanced with anti-mouse γ serum, showed the usual frequency of double PFCs ($24/10^6$ cells or 1.9% of total number of PFCs). Enhancement of PFCs from the same cell populations with specific anti-allotype sera resulted in a sharp reduction of the frequency of doubles. Micromanipulation of double PFC from the initial anti-μ anti-γ enhanced medium to a second medium similarly enhanced showed that 53% of micromainpulated cells continued to lyse both indicators. When double PFCs were transfered from the anti-μ + anti-γ medium to a second medium enhanced with anti-μ + anti-DBA/2 allotype (Ig-1^c), only 7% of these PFCs produced a double lysis (Table 5). It appears from these preliminary experiments that in the vast majority of double PFCs both chromosomes were active. This would indicate that, in a part

Table 3. Culture of individual double anti-TNP and anti-SRBC PFC (from primarily immunized Swiss mice with TNP-SRBC) or double anti-TNP and anti-PC PFC (from secondarily immunized mice with TNP-KLH and PC-KLH). After 48 hrs the cells of each microculture were plated with both indicator red cells (native SRBC and TNP-PRBC or PC-PRBC and TNP-SRBC respectively) and the daughter PFC developed by the direct or enhanced method.

	Double PFC anti-TNP and anti-SRBC	Double PFC anti-TNP and anti-PC
N° and specificity of individually cultured parental double PFC	71	64
N° and % of parental PFC that yielded a progeny	19	12 (19%)
N° of parental PFC that generated daughter PFC (→ N° of daughters) of identical specificity	3 → 3 doubles	0
N° of parental PFC that generated variant daughter PFC (→ N° of daughters and specificity)	16 → 17 $\bar{\alpha}$ - SRBC 16 → 13 $\bar{\alpha}$ - TNP 16 → 0 doubles	12 → 18 $\bar{\alpha}$ - TNP 12 → 14 $\bar{\alpha}$ - PC 12 → 3 doubles

Table 4. Secondary response of (C57B1xDBA/2)F_1 mice immunized with KLH-TNP and KLH-PC. The spleen cells were collected on the 4th-6th day after the 2nd injection, plated with TNP-PRBC and PC-SRBC and enhanced by the mentioned antisera.

Enhancing antisera anti-μ +	PFC/10^6 cells							
			Doubles (Exp. N°)					
	anti-TNP	anti-PC	1	2	3	4	5	6
Alone	9 (± 8)[x]	10 (± 10)	0	0	0	0	0	0
Anti-γ	676 (±439)	1106 (±652)	10	120	0	23	0	50
anti-DBA/2 ($Ig-1^c$)	401 (±369)	1030 (±404)	0	60	0	0	0	12
anti-C57B1 ($Ig-1^b$)	61 (± 29)	388 (±160)	6	0	0	0	0	6

[x]average (± S.E.) out of 6 experiments

Table 5. Micromanipulation of PFC from (C57BlxDBA/2)F_1 mice secondarily immunized with SRBC and PRBC. The cells were plated with both SRBC and PRBC and enhanced with the mentioned antisera.

First Medium	Second Medium			
	PFC enhanced with anti-μ + anti-γ (controls)			
	TOTAL	DOUBLES	SIMPLES	NONE
	68	35	14	19
	(100%)	(53%)	(20%)	(27%)
PFC enhanced with anti-μ + anti-γ	PFC enhanced with anti-μ + anti-DBA/2 (Ig-1^c)			
	TOTAL	DOUBLES	SIMPLES	NONE
	55	4	34	17
	(100%)	(7%)	(62%)	(31%)

of the total PFCs appearing in these double responses (300 to 1500/10^6 spleen cells, Table 1), both chromosomes are functional and active. Yet, persistence of some double PFCs in both sets of experiments (Tables 4, 5) does not exclude that a few cells the same chromosome may code for both antibody molecules.

III. ANTIGEN DRIVEN INTRACLONAL SPECIFICITY VARIATION

The use of individual cell cultures disclosed another important feature of the progeny of single PFCs from doubly immunized animals. It was expected that once double PFCs divided and generated monospecific PFCs, the progeny of each would maintain the same specificity and constitute one clone. Microculture of single monospecific PFCs from doubly immunized animals, however, gave rise to two types of progeny. One part of parental PFCs either anti-TNP or anti-PC (about one half) generated daughter PFCs of identical specificity in a typical clonal manner (Table 6). However, the other half of parental PFCs generated also monospecific daughters but of either one or other specificity, in a well balanced fashion (Table 6). Thus, a high rate intraclonal specificity variation occurs in the progeny of PFC from animals immunized with two antigens. Some degree of intraclonal variation spontaneously occurs among individually cultured PFC from animals immunized with one antigen (Cunningham and Fordham, 1974; Liacopoulos et al., 1976). This does not exceed 2-5 PFCs out of 100 progeny-yielding PFCs (0.02-0.05). In contrast, the variation observed in the progeny of PFCs from animals immunized with two antigens was 0.55 to 0.58 (Table 6). Furthermore, when one of the two immunizing antigens was lacking for 48 hr from the microculture, generation of PFCs specific for this antigen stopped and the rate of variation decreased to the background levels (0.062) (Couderc et al., 1979). A similar vanishing of intraclonal variation was observed when one of the two antigens added was given as a tolerogen and the other as an immunogen. In control cultures of spleen cells stimulated with TNP-SRBC, the usual rate of variation was observed in the progeny of both anti-TNP and anti-SRBC PFCs (0.69 and 0.64, respectively; Table 7). In experimental cultures of cells made tolerant of TNP epitope and then challenged with TNP-SRBC, the few remaining anti-TNP PFCs mostly did not divide and did not show any variation whereas anti-SRBC PFCs yielded a normal rate of anti-SRBC progeny but very few variants, the rate of intraclonal variation being reduced to only 0.14 (Table 7) (Bleux et al., 1980).

These experiments showing that both lack and excess of antigen result in the disappearance of intraclonal variation clearly indicate that this specificity variation is antigen driven. They further suggest that one cell could be made tolerant of an antigen and still react to another antigen, i.e., to present some kind of compartmentation which allows the inactivation of one specific pathway in parallel with activation of another pathway leading to specific antibody production. Such a compartmentation was shown to occur in mouse SAMM-368 plasmocytoma, which secretes IgA and IgG_{2b} bearing different k chains and not sharing idiotypic determinants. These Ig molecules are segregated just as completely in intracytoplasmic molecules as in secreted molecules (McKeever et al., 1980). Unlike hybridoma cells, these plasmocytoma cells do not produce detectable hybrid molecules. Although not proven, it is probable that double PFCs arising after double immunization, also segregate antibody molecules of the two different specificities. The cellular basis of this compartmentation may well be the simultaneous rearrangement of both sets of chromosomes (coding for H and Lk chains) in SAMM-368 cells as in double PFCs. However, whereas SAMM-368 permanently produces both IgA and IgG_{2b} molecules

Table 6. Number and specificity of daughter PFC generated in cultures of individual anti-TNP or anti-PC PFC from mice given a secondary injection of TNP-KLH and PC-KLH. The daughter PFC were developed after plating cells from individual culture wells with a mixture of TNP-PRBC and PC-SRBC.

	Anti-TNP	Anti-PC
N^{o} and specificity of individually cultured parental PFC	96	102
N^{o} and % of parental PFC that yielded a progeny	24 (25%)	22 (21, 6%)
N^{o} of parental PFC that generated daughters (⟶ N^{o}) of identical specificity	10 ⟶ 16 (0.42)	10 ⟶ 18 (0.45)
N^{o} of parental PFC that generated variant daughters (⟶ N^{o} of daughers and specificity	14 ⟶ 19 $\bar{\alpha}$ - TNP 14 ⟶ 16 $\bar{\alpha}$ - PC 14 ⟶ 2 doubles (0.58)	12 ⟶ 18 $\bar{\alpha}$ - PC 12 ⟶ 22 $\bar{\alpha}$ - TNP 12 ⟶ 0 doubles (0.55)

Table 7. Number and specificity of daughter PFC generated in cultures of individual PFC from cells immunized "in vitro" with TNP-SRBC or cells first incubated "in vitro" with TNP_{51}-HGG (1 μ g/ml for 24 hrs) and then immunized with TNP-SRBC.

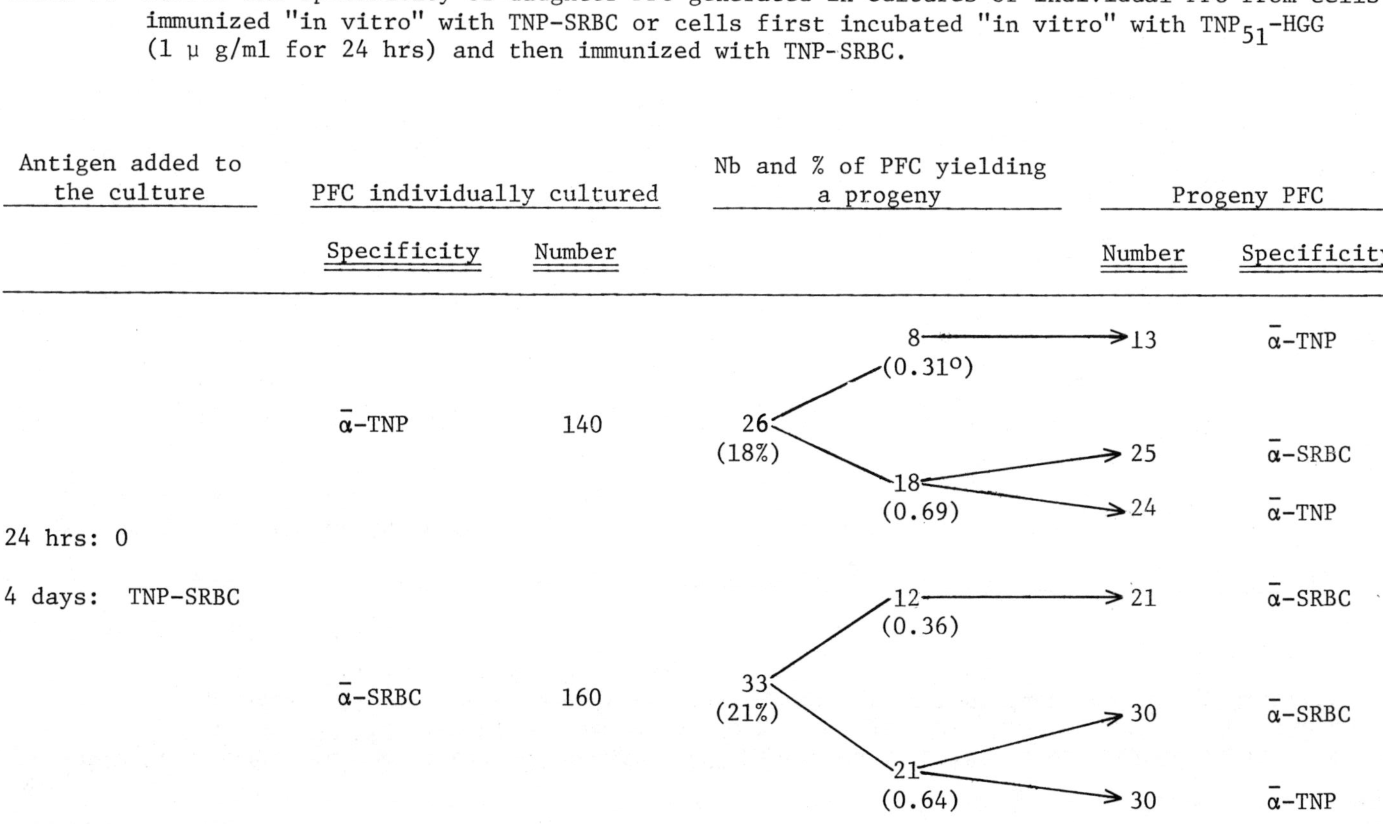

Antigen added to the culture	PFC individually cultured		Nb and % of PFC yielding a progeny		Progeny PFC	
	Specificity	Number			Number	Specificity
	$\bar{\alpha}$-TNP	140	26 (18%)	8 (0.31o)	13	$\bar{\alpha}$-TNP
				18 (0.69)	25	$\bar{\alpha}$-SRBC
24 hrs: 0					24	$\bar{\alpha}$-TNP
4 days: TNP-SRBC	$\bar{\alpha}$-SRBC	160	33 (21%)	12 (0.36)	21	$\bar{\alpha}$-SRBC
				21 (0.64)	30	$\bar{\alpha}$-SRBC
					30	$\bar{\alpha}$-TNP

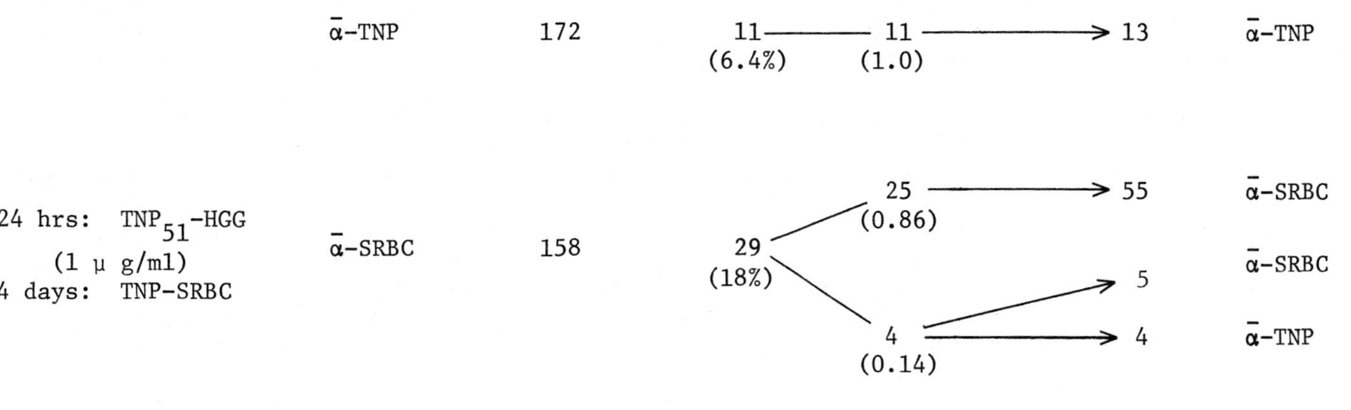

	$\bar{\alpha}$-TNP	172	11 (6.4%)	11 (1.0)	→ 13	$\bar{\alpha}$-TNP
24 hrs: TNP_{51}-HGG (1 μ g/ml) 4 days: TNP-SRBC	$\bar{\alpha}$-SRBC	158	29 (18%)	25 (0.86)	→ 55	$\bar{\alpha}$-SRBC
				4 (0.14)	→ 5	$\bar{\alpha}$-SRBC
					→ 4	$\bar{\alpha}$-TNP

° fraction of invariant versus variant parental PFC

Table 8. Spontaneous PFC either anti-SRBC or anti-HoRBC, or immune (of 2nd or 3rd day) anti-TNP-HoRBC individually cultured with the reciprocal antigen for 34 or 48 hrs. Number and specificity of daughter PFC revealed after plating with native SRBC and TNP-PRBC.

	Spontaneous anti-TNP PFC cultured with SRBC for 24 hrs	Spontaneous PFC cultured with the reciprocal antigen for 48 hrs		Immune anti-TNP-HoRBC cultured with SRBC for 48 hrs
		Anti-TNP PFC cultured + SRBC	Anti-SRBC PFC cultured with TNP-HoRBC	
N^o and specificity of parental PFC	112	104	92	352
N^o and % of parental PFC that yielded a progeny	20 (18%)	23 (22%)	17 (18.4%)	81 (23%)
N^o of parental PFC that generated daughters ($\longrightarrow N^o$) of identical specificity	20 → 43 $\bar{\alpha}$-TNP	12 → 25 $\bar{\alpha}$-TNP	6 → 10 $\bar{\alpha}$-SRBC	37 → 56 $\bar{\alpha}$-TNP
N^o of parental PFC that generated variant daughters ($\rightarrow N^o$ of daughters and specificity	0 $\bar{\alpha}$-SRBC	11 → 11 $\bar{\alpha}$-TNP 11 → 16 $\bar{\alpha}$-SRBC	11 → 10 $\bar{\alpha}$-SRBC 11 → 16 $\bar{\alpha}$-TNP	44 → 36 $\bar{\alpha}$-TNP 44 → 45 $\bar{\alpha}$-SRBC 44 → 10 doubles

Total progeny	43 $\bar{\alpha}$-TNP 0 $\bar{\alpha}$-SRBC	36 $\bar{\alpha}$-TNP 16 $\bar{\alpha}$-SRBC	20 $\bar{\alpha}$-SRBC 16 $\bar{\alpha}$-TNP	96 $\bar{\alpha}$-TNP 45 $\bar{\alpha}$-SRBC 10 doubles

during successive divisions, normal double PFCs cease to produce one of these antibodies after division. This suggests that some kind of intracellular regulation is opposed to simultaneous double production, and the low percentage of detected double cells (0.5 to 2%) would be cells which escaped this regulation. Indeed, the proportion of PFCs giving rise to a variant progeny is much higher than the proportion of double cells (12-14% out of total microcultured PFCs, 55-58% of progeny yielding PFCs; Table 6). Moreover, the existence of this intraclonal variation, which continues at the same rate for at least 15 days after double immunization (Liacopoulos et al., 1979), suggests that the intracellular regulation preventing simultaneous production of two antibody molecules most likely does not stem from permanent inactivation of one chromosome but from a temporary blockage of its expression at an unknown level since in the next cell generation the previously silent chromosome is now expressed.

IV. CONVERSION FROM ONE TO ANOTHER V GENE EXPRESSION

The aforementioned need of the continuous presence of both antigens for persistence of intraclonal variation (Couderc et al., 1979) would indicate that early PFCs still bear receptors unrelated to the antibody produced. The question arose as to whether spontaneous or early immune PFCs secreting antibody of a given specificity can generate, when properly stimulated with an unrelated antigen, daughter PFCs producing antibody specific to the latter antigen. Spontaneous PFCs either anti-SRBC or anti-TNP individually cultured with the corresponding antigen for 24 to 48 hr did not produce any variant progeny. They also did not yield variant progeny when cultured with the alternate antigen for 24 hr. However, when spontaneous PFCs of these specificities were individually cultured with the reciprocal antigen for 48 hr, an large proportion of the parental PFCs yielded a variant progeny comprising daughter PFCs of the parental specificity as well as daughters specific to the antigen introduced into the culture. The same result was obtained when anti-TNP PFCs taken on the 3rd day after *in vivo* immunization with TNP-horse RBC were individually cultured with SRBC (Table 8). In all, about one-third of the daughter PFCs produced antibody specific to what was for them a new antigen, the other two-thirds maintaining the specificity of the parental PFCs (Couderc et al., 1979). This induction of the synthesis of a new antibody specificity by stimulation of PFCs with a different antigen indicates that early PFCs producing antibody of a given specificity, should be still capable recognizing unrelated antigens and reacting to them.

Another type of conversion of V gene expression was studied in this laboratory. Miller et al. (1975) observed that at the early stage of the response to horse radish peroxydase (PO) some cells contain both specific anti-PO antibody and immunoglobulins without detectable anti-PO activity in different areas of their cytoplasm. As this observation suggested that some nonspecific Ig producing cells, appearing after stimulation with a given antigen, might be precursors of specific antibody producing cells, we undertook a series of experiments using the individual cell culture method to study this possibility. Lymph node cells, taken on the 6th day after injection of 250 μg of PO incorporated in Freund's complete adjuvant in hind footpads, were plated with either mouse anti-Ig coated SRBC (for developing reverse hemolytic plaques), PO-SRBC, or native SRBC. They were found to contain 872 ± 474 Ig-secreting PFC/10^6 cells as compared to only 4 ± 6 anti-PO and 1 ± 3 anti-native SRBC PFC/10^6 cells (Antoine et al., 1979). Nonspecific Ig forming cells were micromanipulated from reverse plaques and individually cultured for 48 hr in the presence of PO. As shown in Table 9, out of 502 microcultured cells, 120 (23.9%) yielded anti-PO progeny (519 daughter PFCs). Thus, a fourth of nonspecific Ig forming cells generated anti-PO producing daughters whereas the frequency of such cells among the parental cells was only 1:200 (4 ± 6 vs 872 ± 474). The specificity of the duaghter cells was tested by adding soluble PO into the hemolytic plaque formation cell suspension. Under such conditions, only two parental cells (out of 119; 1.7%) generated three daughters lysing PO-SRBC indicators (Table 9). Similarly, if nonspecific Ig forming cells are cultured in the absence of PO, only two (out of 127; 1.6%) generated eight daughter PFCs lysing PO-SRBC indicators. This finding that the presence of immunizing antigen is required would mean that, during the microculture period (6th to 8th day after immunization) pluripotential Ig forming cells are converted to specific anti-PO forming cells and for this conversion, continuous presence of antigen is necessary. The same period corresponds also to the *in vivo* appearance in significant numbers of anti-PO producing cells (8th to 9th day after immunization) (Avrameas et al., 1976).

As in the previously mentioned experiments (Table 8), the experiments analyzed above (Table 9) clearly show that the product of the daughter cells express an entirely different specificity since the antibody or Ig produced by parental cells did not show any reactivity to the antigen (PO) reacting with daughter cells. This lack of cross reactivity between parental and daughter PFCs makes it unlikely that this conversion could stem from a reassociation of the same H chains with different L chains. It is more probable that this conversion would imply more profound changes in the genome, either activation of second chromosome or rearrangement of new V genes with C genes.

Table 9. Generation of daughter anti-PO PFC during 48 hrs individual culture of non anti-PO immunoglobulin producing parental PFC+, appearing on the 6th day after immunization of mice with PO in Freund's complete adjuvant.

N^{o} of parental microcultured PFC	Presence (+) or not (-) of PO in the microculture	Presence (+) or not (-) of soluble PO in PFC developing medium	Total N^{o} of anti-PO-SRBC after 48 hrs microculture	Total N^{o} of anti PO-SRBC PFC after 48 hrs microculture
502	+	-	120 (23.7)	519
119	+	+	2 (1.7)	3
127	-	-	2 (1.6)	8

+Developed by the reverse method using anti-mouse Ig coated SRBC.

The possibility that the second chromosome is active in double PFCs was previously examined (Tables 4, 5) and it was found that in the majority of such cells examined from F_1 hybrids mice both chromosomes were active. In some cells, however, the same specific anti-allotype serum enhanced a double lytic activity thus suggesting that both antibodies were coded by the same chromosome. A more precise answer to this question was sought for, by testing with the individual PFC culture method, cells from animals immunized with three non-cross-reacting erythrocytes, namely, human RBC, SRBC, and PRBC, where morphological differences enabled distinction under the microscope. After immunization of this type, high numbers of monospecific and bispecific PFCs were detected (Table 10), and some hemolytic plaques in which all three erythrocyte types were lysed. However, in the majority of the latter plaques the lysis was performed not by single cells but by clumps of cells which suggested that the lysis could be due to a clone of dividing cells. In order to simplify the experimental procedure, single anti-HuRBC PFCs were micromanipulated, individually cultured with the three types of erythrocytes for 48 hr, and the daughter cells were plated with SRBC and PRBC in order to see whether PFCs specific to one of the immunizing antigens (Hu-RBC) could generate daughter cells producing antibodies specific to the two other antigens (SRBC and PRBC).

The results of these experiments reported in Table 10 showed, first, that the usual proportion of parental PFCs (22%) yielded a progeny and, second, that this progeny included both anti-SRBC and anti-PRBC daughter monospecific PFCs as well as double PFCs lysing both these indicators. Therefore, if an anti-HuRBC PFC is able to generate daughter PFCs expressing the one, the other or both specificities of the other two immunogens, at least two of the three specific reactivities must be coded by the same chromosome.

V. DISCUSSION

It must be stressed that this type of work, i.e., examining the activity of rapidly dividing and differentiating cells, must only be done at the single cell level and no other method than PFC is presently available for detecting and characterizing antibody produced by individual cells. Repeated attempts to produce double hybridomas by fusing myeloma cells with mouse spleen cells taken and the period where double cells appear (3rd to 6th day after primary or secondary immunization) yielded many functional hybridomas some of which initially produced the two corresponding antibodies, but this double production ceased a few days after cloning and only the production of one antibody type was stabilized.

Table 10. Direct spleen PFC numbers (/10^6 cells) detected on the 4th day after simultaneous immunization of Swiss mice with sheep (S), pigeon (P), and human (Hu) red blood cells.

	Lysed Red Blood Cells						
	S	P	Hu	S+P	S+Hu	P+Hu	S + P + Hu
Direct PFC	1956	2162	521	261	40	26	11
(/10^C Cells)	±676	±1256	±208	±101	± 8	± 9	± 5

Number and specificity of daughter PFC generated in cultures of individual parental PFC expressing only anti-HuTBC activity.

Total individually cultured anti-HuRBC PFC	163
N^o and % of parental PFC which yielded a progeny	36 (22%)
N^o and specificity of daughter PFC	15 anti-SRBC 23 anti-PRBC
Total progeny	37 doubles
	75

Therefore, it could be asked whether hemolysis producing material secreted by PFCs really is conventional antibody. This problem was extensively studied when the method of detecting individual antibody producing cells by the local hemolytic plaque method was discovered (Jerne and Nording, 1963) and subsequent work amply confirmed the conventional nature of antibody secreted by PFCs. The present work dealing with cells producing a lysis of two indicator red cells with different specificity show that lytic material secreted by these cells should be considered as conventional antibodies since (1) each lytic activity is solely inhibited by its specific inhibitor, (2) secondary double cells need enhancing anti-mouse IgG antibody to become lytic, (3) the same enhancement is produced by specific anti-mouse allotype sera, and (4) specific anti-mouse μ serum inhibits direct plaques.

The present experimental results are not contradictory with findings reported by others in the past. The absolute predominance of monospecific PFCs, after immunization with two antigens (at least 98%), was again found in the present studies. The appearance of a few doubles after such immunization was also repeatedly reported in the past (cf. Makela and Cross, 1970) and especially by Attardi et al., (1964) who demonstrated that these cells produced two unrelated antibodies. Similarly, intraclonal specificity variation could be deduced from experiments of White (1958) and Green (1968) who found that, after immunization with two unrelated proteins, cells specifically stained with each antigen (labelled with fluorescein and rhodamine) were randomly intermixed in lymphoid organs. Negative results reported by other authors during the last 15 years could be explained by differences in experimental methods; for example, they were not looked for at a suitable time after double immunization (Couderc et al., 1973). Resistance to admitting the existence of cells producing two antibody molecules would probably stem from incompatibility of the existence of such double cells with current theories on the generation of antibody diversity. Indeed both theories, calling upon either a germinal or somatic orgin of antibody diversity, propose that B cells are unipotential, the commitment to the unique specificity occuring either by permanent rearrangement of a given V gene with C gene at the stage of the pre-B cells or by occurrence of one mutation on a rearranged V gene during the early period of ontogeny. The clonal development of these specified B cells constitutes the antibody repertoire of the animal. This clonal expansion of specified B cells received experimental support not only by findings that the vast majority of antibody producing cells as well as induced or spontaneous myelomas (with some exceptions in mouse Morse et al., 1976 and in human myelomas, Bouvet et al., 1974, Hopper, 1977) produce a single antibody species, but also by findings that elimination of cells of a given specificity from a normal lymphoid cell population obliterates the capacity of this population to produce an immune response of the same specificity without interfering with development of unrelated responses (Wigzell and Makela, 1970). However, if such populations are stimulated with the same antigen and an adjuvant, they rapidly recover lacking reactivity (Prunet et al., 1978).

More recent experiments on Ig receptors of B cells showed an unexpectedly high frequency of normal B cells bearing receptors for certain antigens individually tested (cf. Sigal and Klinman, 1978); and when receptors for two different antigens were looked for, cells simultaneously binding two different antigens were found in significant numbers (DeLuca et al., 1975). Furthermore, the same authors observed that when an antigen induces capping of surface Ig receptors, all these receptors co-cap, including receptors for other specificities. This observation explains why Raff et al. (1973) found that when polymerized flagellin specific receptors are capped, all surface Ig molecules are drawn along into the same cap. On the basis of the frequency of B cells bearing receptors for individual antigens, Miller (1977) calculated that each B cell could express up to 10^2 different antigen receptors distributed among the 1 to $5x10^5$ total Ig molecules of its membrane (Rabellino et al., 1971). There is no reason to assume that the 10^2 different specific receptors are equally distributed in all cells. On the contrary, Klinman and Press's (1975) demonstration of the existence of "predominant", "sporadic", and "rare" clonotypes, implies a nonrandom distribution, especially as this distribution changes during the life of the animal. Instead of attributing this distribution to the relative frequency of specific B cells, the relative frequency (or density) of receptors of a given specificity on individual cells, could equally well account for unequal numbers of B cell precursors responding to different antigens. It could be assumed indeed that there is a threshold frequency of receptors of a given specificity that enables triggering of the cell by the corresponding antigen, or its retention on the immunoabsorbent columns. Such a threshold frequency of specific receptors on individual cells would limit the number of precursor cells triggered by a given antigen. After stimulation with two antigens, the cells that, by chance or after preimmunization, bore both specific receptors in sufficient frequency, would be triggered, then differentiate into double cells and/or give rise to intraclonal specificity variation presenting clones.

Although such a model could account for cellular events consecutive to the introduction of antigen, it is not supported by current data on events at the DNA level during the maturation of B cells. There are two major limitations. Firstly, of the two chromosomes, only one was found to be rearranged in myeloma cells (Brack et al., 1978). More recent work showed that in other myeloma, lymphoma, and especially normal spleen cells, both chromosomes undergo rearrangement (Hurwitz et al., 1980) but rearrangement of the second chromosome would not be productive (Van Ness et al., 1981). However, as mentioned previously in most double cells (Table 5), both chromosomes are active and the phenomenon of intraclonal specificity variation could most likely explained by temporary blockage of the expression of the second rearranged chromosome. It should be stressed, therefore, that as these events occur in a very restricted number of cells (about $100/10^6$ spleen cells), they remain undetected by present procedures of molecular biology. The only permanent line producing two types of Ig molecules (SAMM 368 myeloma; Morse et al., 1977) has not as yet been analyzed at the DNA level.

The second limitation is that only one V gene joins J and C genes in the rearranged chromosome (Brack et al., 1978), and this type of rearrangement is considered as occuring at the pre-B cell level and imposing monopotentiality of B cells. Existence of double cells is incompatible with such a view, even if it could be admitted that in some cells both chromosomes are functionally rearranged. Actually, since the choice of the V gene that is to be rearranged is obviously random in each chromosome, simultaneous rearrangement of the two V genes corresponding to the pair of antigens chosen by the experimenter in the same cell should be extremely rare. Yet, during this work we tested more than ten different antigen pairs and in all cases we found the usual low percentage of doubles (0.5 to 2%). Similarly conversion from one specificity to another (Tables 8, 9) should never be observed. An insight as to what might happen at the level of antibody genes was suggested by experiments showing that, in a low proportion of doubles from F_1 hybrid mice, both specific reactivities bear the same allotype (Tables 4, 5) and especially after immunization with three antigens where a cell expresses one of the three specificities, it can generate daughters expressing the other two specificities (Table 10). These findings suggest that each chromosome could code for two different specificities, either simultaneously or, more probably, successively by transcription of one V-C gene and translation of the second specificity from a stable mRNA. According to this view the multiple antigen receptors found on the surface of resting B cells could be generated from successive rearrangements of different V genes with the same C gene and their subsequent transcription. Taking into account the existence of stable mRNA and the rate of turnover of membrane immunoglobulins, it is conceivable that antigen receptors could persist longer on the surface of cells than one given V-C rearrangement of antibody genes.

An additional assumption, necessary for fully explaining above findings, is that once a V gene is arranged and expressed, it is not deleted but remains available for a new arrangement and expression. This also implies that the second rearrangement of an already expressed V gene is dictated by or linked in some way with the corresponding surface receptor. A successive arrangement and expression occurs at the level of C genes and, although Yaoita and Honjo (1980) found that in myeloma cells left-hand C genes to the expressed one are deleted, three independent reports described reexpression of already expressed C genes in another myeloma (Radbruch et al., 1980) and in normal cells (Ventura et al., 1978; Ault and Towle, 1981).

As C gene switches appear to be influenced or dictated by extra and/or intracellular regulation, successive V gene rearrangement and expression could also be regulated by intracellular events consecutive to antigen binding to specific receptors alone (signal 1) or in conjunction with the mitogenic signal 2, about which nothing is known at the present time. In order to find out the molecular basis of this intracellular regulation, studies in small clones of cells at several points during their expansion before and after antigen stimulation should be made workable. While waiting for this possibility, one has to take into account what is observed at the whole cell level.

VI. REFERENCES

Antoine, J. C., Bleux, C., Avrameas, S., and Liacopoulos, P. (1979). Specific antibody-secreting cells generated from cells producing immunoglobulins without detectable antibody function. *Nature*, 277, 218-219.

Attardi, G., Cohn, M., Hobirata, K., and Lennox, M. (1964). Antibody formation by rabbit lymph node cells. I. Single cells responses to several antigens. *J. Immunol.*, 92, 335-345.

Ault, K. A. and Towle, M. (1981). Human B lymphocyte subsets. I. IgG-bearing B cell response to pokweed mitogen. *J. Exp. Med.*, 153, 339-351.

Avrameas, S., Antoine, J. C., Ternynck, T., and Petit, C. (1976). Development of immunoglobulin and antibody-forming cells in different stages of the immune response. *Ann. Immunol.* (Institut Pasteur), 127C, 551-571.

Bleux, C., Desaymard, C., Couderc, J., and Liacopoulos, P. (1980). Immunity and tolerance induced by two antigens in individual B cells. *Ann. Immunol.* (Institut Pasteur), 131D, 173-185.

Bouvet, J. P., Buffe, D., Oriol., R., and Liacopoulos, P. (1974). Two myeloma globulins, $IgG_{1\kappa}$ and $IgG_{1\lambda}$ from a single patient (Im). II. Their common cellular origin as revealed by immunofluorescence studies. *Immunology*, 37, 1095-1101.

Brack, C., Hirama, M., Lenhard-Schuller, R., and Tonegawa, S. (1978). A complete immunoglobulin gene is created by somatic recombination. *Cell*, 15, 1-14.

Couderc, J., Birrien, J. L., Oriol, R., Bleux, C., and Liacopoulos, P. (1975). Bispecific cells among IgM and IgG producers during the early phase of primary and secondary responses. *Eur. J. Immunol.*, 5, 140-147.

Couderc, J., Bleux, C., Birrien, J. L., and Liacopoulos, P. (1973). Transcient appearance of cells secreting antibodies of different specificities after immunization of mice with trinitrophenylated erythrocytes. *J. Immunol.*, 111, 1155-1163.

Couderc, J., Bleux, C., Ventura, M., and Liacopoulos, P. (1979). Single mouse cells producing two antibody molecules and giving rise to antigen driven intraclonal variation after immunization with two unrelated antigens. *J. Immunol.*, 123, 173-181.

Cunningham, A. J. and Fordham, S. A. (1974). Antibody cell daughters can produce antibody of different specificities. *Nature*, 250, 669-671.

DeLuca, D., Miller, A., and Sercarz, E. (1975). Antigen binding to lymphoid cells from unimmunized mice. IV. Shedding and re-appearance of multiple antigen binding Ig receptors of T- and B-lymphocytes. *Cell. Immunol.*, 18, 286-303.

Green, I. (1968). Distribution of antibody-forming cells of different specificities in the lymph nodes and spleens of guinea pigs. *J. Exp. Med.*, 128, 729-749.

Hopper, J. E. (1977). Comparative studies on monotypic IgMλ and IgGκ from a individual patient. IV. Immunofluorescent evidence for a common clonal synthesis. *Blood*, 50, 203-211.

Hurwitz, J. L., Coleclough, C., and Cebra, J. J. (1980). C_H-gene rearrangement in IgM bearing B-cells and in the normal splenic DNA component of hybridomas making different isotypes of anti-body. *Cell*, 22, 349-359.

Jerne, N. K. and Nordin, A. A. (1963). Plaque formation in agar by single antibody producing cells. *Science*, 140, 405.

Klinman, N. R. and Press, J. L. (1975). The B cell specificity repertoire: Its relationship to definable subpopulations. *Transplant. Rev.*, 24, 41-83.

Liacopoulos, P., Couderc, J., and Bleux, C. (1976). Evidence for multipotentiality of antibody synthesizing cells. *Ann. Immunol.* (Institut Pasteur), 127C, 519-530.

Mäkelä, O. and Cross, A. M. (1970). The diversity and specialization of immunocytes. *Progr. Allergy*, 14, 145-207.

McKeever, P. E., Neiders, M. E., Nero, G. B., and Asofsky, R. (1980). Murine plasma cells secreting more than one class of immunoglobulin. VII. Analysis of the IgG_{2B} and IgA precursors within the cytoplasm of spontaneous myeloma SAMM 368 in culture shows segregation of heavy chains. *J. Immunol.*, 124, 541-547.

Miller, A. (1977). A critical examination of the numerology of antigen binding cells: Evidence for multiple receptor specifi-cities on single cells. *Ann. Immunol.* (Institut Pasteur) 128C, 611-620.

Miller, H. R. P., Ternynck, T., and Avrameas, S. (1975). Synthesis of antibody and immunoglobulins without detectable antibody function in cells responding to horseradish peroxidase. *J. Immunol.*, 114, 626-629.

Morse, H. C., Neiders, M. E., Lieberman, R., Lawton, A. R., III, and Asofsky, R. (1977). Murine plasma cells secreting more than one class of immunoglobulin. II. SAMM 368-A plasmocytoma secreting $IgG_{2B}κ$ and IgAκ immunoglobulins which do not share idiotypic determinants. *J. Immunol.*, 118, 1682-1689.

Morse, H. C., Pumphrey, J. G., Potter, M., and Asofsky, R. (1976). Murine plasma cells secreting more than one class of immunoglobulin heavy chain. Frequency of two or more M-components in ascitic fluids from 788 primary plasmocytomas. *J. Immunol.*, 117, 541-547.

Prunet, J., Birrien, J. L., Panijel, J., and Liacopoulos, P. (1978). On the mechanism of early recovery of specifically depleted lymphoid cell populations by non-specific activation of T cells. *Cell. Immunol.*, 37, 151-161.

Raff, M. C., Feldmann, M., and de Petris, S. (1973). Monospecificity of bone marrow-derived lymphocytes. *J. Exp. Med.*, 137, 1024-1030.

Rabellino, E., Colon, S., Grey, H. M., and Unanue, E. R. (1971). Immunoglobulins on the surface of lymphocytes. I. Distribution and quantitation. *J. Exp. Med.*, 133, 156-167.

Radbruch, A., Liesegang, B., and Rajewsky, K. (1980). Isolation, of variants of mouse meyloma X63 that express changed immunoglobulin class. *Proc. Natl. Acad. Sci. USA*, 77, 2909-2913.

Schilling, J., Cleringer, B., Davie, J. M., and Hood, L. (1980). Aminoacid sequence of homogeneous antibodies to dextran and DNA rearrangements in heavy chain V-region gene segments. *Nature*, 283, 35-40.

Sigal, N. H. and Klinman, N. R. (1978). The B-cell clonotype repertoire. *Adv. Immunol.*, 26, 255-337.

Van Ness, B., Perry, R., and Weigert, M. (1981). Rearrangements and expression of immunoglobulin κ genes. *J. Supramol. Struct. and Cell. Biochem.*, suppl. 5, P. 18 (Abstract).

Varga, J. M., Koningsberg, W. H., and Richards, F. F. (1973). Antibodies with multiple binding functions. Induction of single immunoglobulin species by structurally dissimilar haptans. *Proc. Natl. Acad. Sci. USA*, 70, 3269-3274.

Ventura, M., Bleux, C., Crepin, Y., and Liacopoulos, P. (1978). Ig-isotype diversity generated in antibody forming cell of the mouse. *J. Immunol.*, 121, 817-822.

White, R. G. (1958). Antibody production by single cells. *Nature*, 182, 1383-1384.

Wigzell, H. and Mäkelä, O. (1970). Separation of normal and immune lymphoid cells by antigen-coated columns. Antigen binding characteristics of membrane antibodies as analyzed by hapten-protein antigens. *J. Exp. Med.*, 132, 110-126.

Yaoita, Y. and Honjo, T. (1980). Deletion of immunoglobulin heavy chain genes from expressed allelic chromosome. *Nature*, 286, 850-853.

INDEX